INTELLIGENT SENSOR NODE-BASED SYSTEMS

Applications in Engineering and Science

INTELLIGENT SENSOR NODE-BASED SYSTEMS

Applications in Engineering and Science

Edited by

Anamika Ahirwar, PhD

Piyush Kumar Shukla, PhD

Prashant Kumar Shukla, PhD

Ruby Bhatt, PhD

AAP | APPLE ACADEMIC PRESS

First edition published 2024

Apple Academic Press Inc.
1265 Goldenrod Circle, NE,
Palm Bay, FL 32905 USA

760 Laurentian Drive, Unit 19,
Burlington, ON L7N 0A4, CANADA

CRC Press
2385 NW Executive Center Drive,
Suite 320, Boca Raton FL 33431

4 Park Square, Milton Park,
Abingdon, Oxon, OX14 4RN UK

Library and Archives Canada Cataloguing in Publication

Title: Intelligent sensor node-based systems : applications in engineering and science / edited by Anamika Ahirwar, PhD, Piyush Kumar Shukla, PhD, Prashant Kumar Shukla, PhD, Ruby Bhatt, PhD.
Names: Ahirwar, Anamika, editor. | Shukla, Piyush Kumar, 1976- editor. | Shukla, Prashant Kumar, editor. | Bhatt, Ruby, editor.
Description: First edition. | Includes bibliographical references and index.
Identifiers: Canadiana (print) 20230451950 | Canadiana (ebook) 20230451993 | ISBN 9781774913239 (hardcover) | ISBN 9781774913246 (softcover) | ISBN 9781003372042 (ebook)
Subjects: LCSH: Wireless sensor networks. | LCSH: Internet of things.
Classification: LCC TK7872.D48 I58 2023 | DDC 006.2/5—dc23

Library of Congress Cataloging-in-Publication Data

CIP data on file with US Library of Congress

ISBN: 978-1-77491-323-9 (hbk)
ISBN: 978-1-77491-324-6 (pbk)
ISBN: 978-1-00337-204-2 (ebk)

About the Editors

Anamika Ahirwar, MCA, PhD
Associate Professor and Head, Computer Science Department,
Compucom Institute of Information Technology and Management,
Jaipur, Rajasthan, India

Anamika Ahirwar, PhD, is working as an Associate Professor and Head of the Computer Science Department in Compucom Institute of Information Technology & Management, Jaipur, Rajasthan, India. She has 18 years of experience in teaching and research. She has published more than 45 research papers in reputed national and international journals and conferences. She has authored and reviewed many books published by national and international publisher. She completed her PhD degree in Computer Applications at RGPV, Bhopal, India. She acquired her master's degree (MCA) from Government Gitanjali Girls PG College, Bhopal (M.P.), India, and completed her BSc (Maths) degree at Government K.R.G. PG Autonomous College, Gwalior, M.P., India. Dr. Ahirwar's research areas include medical imaging, data mining, celestial sound, IoT, and machine learning. She has delivered several expert and guest lectures, attended seminars, and chaired sessions at various IEEE and other international conferences. She is actively serving as a journal reviewer for various publishers, such as IEEE, Springer, John Wiley, and IGI Global and is also as an editorial board member of IEEE conferences and other reputed international journals and conferences. She of the International Association of Engineers (IAENG), Hong Kong. She has published five patents. Dr. Ahirwar has guided many research projects. She has been awarded as a National Eminent Researcher Award, National Eminent Academic Influencer Award, and an Academic ICON Award Shikshak.

Piyush Kumar Shukla, PhD

Piyush Kumar Shukla, PhD, is Associate Professor in the Department of Computer Science and Engineering at the University Institute of Technology-Rajiv Gandhi Proudyogiki Vishwavidyalaya (RGPV, Technological University of Madhya Pradesh), Bhopal, M.P., India, from which he also earned his PhD. He has over 15 years of experience in teaching and research. Dr. Shukla's research interests include white-box cryptography, information security and privacy, cybersecurity, dynamic wireless networks, machine learning, image processing, blockchain, and IoT. He has published several books and book chapters with international publishers. He has also published more than 15 Indian patents and papers in international journals and conferences. He was selected as Best Researcher of the Year–2019 for Outstanding Research Contribution—Network Security, an international research award. He has also successfully completed a NPTEL 12-week course on digital circuits conducted by the Indian Institutes of Technology, Kharagpur, India. He has delivered several expert/guest lectures, attended seminars, and chaired sessions at various IEEE and Springer international conferences. Dr. Shukla is a senior life member of IEEE and a branch counselor of the IEEE Student Branch at RGPV, Bhopal, India. He is also a single point of contact for National Programme on Technology Enhanced Learning (NPTEL), SIH (Smart India Hackathon). He is actively serving as an editorial board member and reviewer for various IEEE, Springer, and IGI Global journals. He has supervised many PhD and MTech dissertations and has served the department Postgraduate In-charge/Coordinator. He has completed a post-poctorate fellowship recently under the Information Security Education and Awareness Project Phase II, funded by the Ministry of Electronics and Information Technology. He earned his MTech in CSE from Samrat Ashok Technological Institute.

Prashant Kumar Shukla, PhD

Assistant Professor (SG) at the JLU School of Engineering & Technolog, Bhopal, India

Prashant Kumar Shukla, PhD, joined as Assistant Professor (SG) at the JLU School of Engineering & Technology, Bhopal, India. He has been in research, teaching, and industry for the past 19 years and is working in the research areas of machine learning, deep learning, computer vision, Internet of Things (IoT), etc. He has applied for 24 patents, of which 23 patents have been published. He has received funding for two research projects. He has published and presented over 20 research papers in various national and international SCI/Web of Science/Scopus-indexed journals and conferences. He has also published book chapters. Dr. Shukla has participated as a mentor in Smart India Hackathon (Hardware Edition) 2018 and secured first position (prize money Rs. 100,000) and also secured second runner-up position (prize money Rs. 50,000) in Smart India Hackathon (Software Edition) 2018. Dr. Shukla has received various awards, including Innovative Teacher Award from the GISR Foundation and The American College of Dubai at Dubai, UAE; Best Researcher from ESN Publications, Tamilnadu, India; a Teacher Innovation Award from ZIIEI, Sri Aurobindo Society, India; and Green ThinkerZ Preeminent Researcher award 2019 from the Green ThinkerZ Society, Chandigarh, India. He has been a contributor to several professional institutions, including IAENG, IACSIT, and SDIWC. He is a member of the Tuning India project, which is co-funded by the Erasmus+ Programme of the European Union. He is a member of over 25 editorial and reviewer boards of national and international research journals. He has attended and organized more than 30 workshops, seminars, conferences, and faculty training programs. He is also associated with two business start-ups. Dr. Shukla holds a PhD in Computer Science and Engineering from Dr K. N. Modi University, Rajasthan, India, and a Master of Engineering from Rajiv Gandhi Proudyogiki Vishwavidyalaya, Bhopal, India.

Ruby Bhatt, PhD
Assistant Professor, Department of Computer Science, Medicaps University, Indore, India

Ruby Bhatt, PhD, is associated with Medicaps University, Indore, India, as Assistant Professor in the Department of Computer Science. She has over 19 years of academic experience at universities and colleges in central India. Her area of interest includes wireless sensor networks, security issues in sensor networks, artificial intelligence, and data mining and data analytics. She has also been working on a fruit fly optimization algorithm. She has attended many national and international conferences and has research paper publications to her credit. She has authored journal and conference papers on security issues in wireless sensor networks and optimization techniques. She has mastered many programming languages, including C / C++, Java, HTML / CSS, JavaScript, and high-performance language for technical calculation, MATLAB. Dr. Bhatt has completed her MPhil in Computer Science from Vikram University, Ujjain, India, and MSc in Computer Science from Rani Durgawati University, Jabalpur, India. She has been awarded her Doctor of Philosophy in Computer Science from the Department of Computer Science and Engineering, Rabindranath Tagore University (RNTU), Bhopal, India.

Contents

Contributors .... *xi*

Abbreviations .... *xv*

Preface .... *xvii*

1. **Smart Sensing and Intelligent Systems: An Overview with Applications in Engineering and Sciences 1**
T. P. Kamatchi and K. Anitha Kumari

2. **IoT-Based Smart Chair for Healthcare Supporting System 39**
Pooja Gupta, Sunita Varma, Neeraj Arya, and Upendra Singh

3. **Smart Farming Using Blockchain Technology: An Indian Perspective 53**
Anjana Pandey, Abhishek Dubey, Bhavesh Shah, and Mustafa Kasarawala

4. **An Intelligent Parenting System: Artificial Mother Monitoring System for Sleeping Infant 71**
Harshita Jain

5. **A Review of Smart Intelligent IoT Network: Technologies and Real-Time Applications 95**
Gajendra Kumar Ahirwar, Ratish Agarwal, and Anjana Pandey

6. **Intelligent University Monitoring System (i-UMS) 117**
Akhilesh A. Waoo and Ashwini A. Waoo

7. **Intelligent Home 133**
Lalit Purohit and Manoj Dhawan

8. **Applications and Challenges of IoT-Based Smart Homes 155**
P. S. Patheja, Yatin Kalra, and Akash Tyagi

9. **Intelligent Security System Based on the Internet of Things (IoT) 177**
Pooja Gupta, Sunita Varma, Neeraj Arya, and Ritesh Bhagel

10. Intelligent Agriculture System **193**
Ashwini A. Waoo and Akhilesh A. Waoo

11. Intelligent and Smart Agriculture System Using Cooperative Approach **211**
Bhupesh Gour and Jay Prakash Maurya

12. Synthesis and Fabrication of a Nanosensor Device for Monitoring Nutrient Levels in Aeroponic Agricultural Farming **229**
D. Gajalakshmi

13. Intelligent Smart Sensor for Cognitive Radio Networks: Comparison, Solution, and Analysis **243**
Yogita Thareja, Kamal Kumar Sharma, and Parulpreet Singh

14. Smart and Ecofriendly Intelligent House Based on IoT and Simulation Using a Cisco Networking Simulator **259**
Ravi Ray Chaudhari, Krishna Kumar Joshi, Neelam Joshi, and Anand Kumar Pandey

15. Different Techniques of Data Fusion in the Internet of Things (IoT) **275**
Harsh Pratap Singh, Bhaskar Singh, and Rashmi Singh

Index ***285***

Contributors

Ratish Agarwal
Associate Professor, University Institute of Technology RGPV, Bhopal, Madhya Pradesh, India

Gajendra Kumar Ahirwar
Research Scholar, University Institute of Technology RGPV, Bhopal, Madhya Pradesh, India

Neeraj Arya
Assistant Professor, Department of Information Technology, Shri G. S. Institute of Technology and Science, Indore, Madhya Pradesh, India

Ritesh Bhagel
Student Department of Information Technology, Shri G. S. Institute of Technology and Science, Indore, Madhya Pradesh, India

Ravi Ray Chaudhari
Department of Computer Engineering, SKNSITS Lonavala, India

Manoj Dhawan
Department of Information Technology, Shri Vaishnav Vidyapeeth Vishwavidyalaya, Indore, Madhya Pradesh, India

Abhishek Dubey
Salalah College of Technology, Oman

D. Gajalakshmi
Assistant Professor (SG), Department of Chemistry, University College of Engineering (A Constituent College of Anna University, Chennai), Villupuram, Tamil Nadu, India

Bhupesh Gour
Department of Computer Science and Engineering, Lakshmi Narain College of Technology, Bhopal, Madhya Pradesh, India

Pooja Gupta
Assistant Professor, Department of Information Technology, Shri G. S. Institute of Technology and Science, Indore, Madhya Pradesh, India

Harshita Jain
Harshita Jain, Asst. Prof., Department of Computer Science and Engineering, SIRT, Bhopal

Krishna Kumar Joshi
Department of Computer Engineering, SITS Lonavala, India

Neelam Joshi
Department of Computer Engineering, SITS Lonavala, India

Yatin Kalra
Student, Bachelor of Technology, School of Computing Sciences and Engineering, VIT Bhopal, Madhya Pradesh, India

T. P. Kamatchi
Lecturer, Department of Computer Networking, PSG Polytechnic College, Tamil Nadu, India

Mustafa Kasarawala
University Institute of Technology RGPV, Bhopal, Madhya Pradesh, India

K. Anitha Kumari
Associate Professor, Department of IT, PSG College of Technology, Tamil Nadu, India

Jay Prakash Maurya
Department of Computer Science and Engineering, Lakshmi Narain College of Technology, Bhopal, Madhya Pradesh, India

Anand Kumar Pandey
Department of Computer Science and Engineering, ITM University Gwalior, India

Anjana Pandey
Associate Professor, UIT RGPV Bhopal, Madhya Pradesh, India

P. S. Patheja
Associate Professor (Sr.), School of Computing Sciences and Engineering, VIT Bhopal, Madhya Pradesh, India

Lalit Purohit
Department of Information Technology, Shri Govindram Seksaria Institute of Technology and Science, Indore, Madhya Pradesh, India

Bhavesh Shah
University Institute of Technology RGPV, Bhopal, Madhya Pradesh, India

Kamal Kumar Sharma
School of Electrical and Electronics Engineering, Lovely Professional University, Punjab, India

Bhaskar Singh
Editor in Chief, Bhopal Hundred News24, Bhopal, Madhya Pradesh, India

Harsh Pratap Singh
Department of Computer Science and Engineering, Sri Satya Sai University of Technology and Medical Science, Sehore, Madhya Pradesh, India

Parulpreet Singh
School of Electrical and Electronics Engineering, Lovely Professional University, Punjab, India

Rashmi Singh
MIS Head, Trident Group, Budhni, Hoshangabad, Madhya Pradesh, India

Upendra Singh
Assistant Professor, IT Department, Shri G. S. Institute of Technology and Science, Indore, Madhya Pradesh, India

Yogita Thareja
School of Electrical and Electronics Engineering, Lovely Professional University, Punjab, India, E-mail: yogita.thareja@vips.edu

Akash Tyagi
Student, Bachelor of Technology, School of Computing Sciences and Engineering, VIT Bhopal, Madhya Pradesh, India

Sunita Varma
Professor and Head, Department of Information Technology, Shri G. S. Institute of Technology and Science, Indore, Madhya Pradesh, India

Akhilesh A. Waoo
Head, Department of CS/IT, AKS University, Satna, Madhya Pradesh, India

Ashwini A. Waoo
Associate Professor, Department of Biotechnology, FLST, AKS University, Satna, Madhya Pradesh, India

Abbreviations

AAE	average absolute error
AI	artificial intelligence
AMQP	advanced message queuing protocol
AoC	agent on-chip
ARE	average relative error
BEMS	building energy management systems
BS	base station
CH	cluster heads
CMS	centralized monitoring station
$CO2$	carbon dioxide
CoAP	constrained application protocol
CR	cognitive radio
CRNs	cognitive radio networks
CSMA/CA	carrier senses multiple accesses with collision avoidance
DS	decision stump
DT	decision table
FCC	Federal Communication Commission
FoV	field of view
GD	gradient descent
GUI	graphical user interface
HER	electronic health record
HOG	histogram of oriented gradients
HRs	human resources
HVAC	heating, ventilating, and air conditioning
IAS	intelligent agriculture system
ICT	information and communications technologies
IDE	integrated development environment
IH	intelligent home
IoT	Internet of Things
ISN	intelligent sensor node
ISS	intelligent surveillance system
ITS	intelligent transportation system
JSON	JavaScript object notation

LBPH	local binary pattern histograms
MAC	medium access control
MEMS	micro-electromechanical systems
MQTT	message queuing telemetry transport
MRD	minimum required difference
MSDF	multi-sensor data fusion
MSs	mobile stations
NFC	near-field communication
NN	neural network
NNge	non-nested generalized exemplars
O-D	origin-destination
PAN	personal area network
PIR	passive infrared
QoL	quality of life
QoS	quality of service
RDF	random decision forest
RFID	radio frequency identification
RNN	random neural network
RS	reporting station
SAW	surface acoustic wave
SEH-WSN	solar energy harvesting wireless sensor network
SU	sensor units/nodes
SVM	support vector machine
TVOC	total volatile organic portion
UAV	unmanned aerial vehicle
VPN	virtual private organization
WBI	web-based instructions
WINGSNET	wireless intelligent GPS-based sensor network
WISAN	wireless intelligent sensor and actuator network
WSN	wireless sensor network
XML	extensible terminology

Preface

This book presents a study on current developments, trends, and latest usage of the trending technology in the areas of wireless sensor networks (WSNs). The impending research work on some WSNs applications, like improvements in the agricultural system, security systems, and university monitoring systems, etc., shows the usefulness of sensor networks. These along with theoretical analysis and factors influencing smart sensing are highlighted in this book. The state-of-the-art Internet of Things (IoT) designed and established as analogous to WSNs is also explored in detail.

IoT doesn't assume a specific communication technology, but clubbing it with wireless expertise makes it a major component useful in both engineering as well as businesses. It also makes a good research topic for young researchers. The collaborative usage with IoT is an addendum to WSN trends. This collaborative effort has resulted in the exponential development of mobile traffic in the last few years because an exceptionally large amount of data exists today, and this data is collected by the wireless network industry.

Advancement in the IoT system, along with its two-way usage with the wireless sensor networks, is closing the fissure between the physical and virtual world. As a result, a hyper-connected society has come into existence where devices are not only used to exchange data but are also smart devices with various capabilities. The devices are more context-aware too, that is, even the devices are getting smarter day by day.

The IoT and wireless communication technology have elevated the processes via learning through interactions. The devices are highly optimized. This has led to the formation of a smart and self-aware planet. The definition of primary needs has changed from only food, shelter, and clothing to energy, mobility, digital society, and the interdisciplinary intermingling between natural and man-made things.

Covering all these topics, the book also discusses smart homes, intelligent sensor-based cognitive radio networks, different techniques for data fusion, synthesis, and fabrication of nano-sensor devices for monitoring nutrient levels, etc. Furthermore, it also investigates the fake user problems in WSNs with a note on the current trends as well as the newer trends to come in the near future.

CHAPTER 1

Smart Sensing and Intelligent Systems: An Overview with Applications in Engineering and Sciences

T. P. KAMATCHI and K. ANITHA KUMARI

Lecturer, Department of Computer Networking, PSG Polytechnic College, Tamil Nadu, India

Associate Professor, Department of IT, PSG College of Technology, Tamil Nadu, India

ABSTRACT

Over the years, the entire world is changing rapidly into a technological world. One of the most promising technologies is the intelligent sensor technology which is now available anywhere and everywhere. The intelligent sensor node (ISN)-based system empowers intellectual surroundings, which in turn offer better services in various public or private sectors such as the military, health sector, home automation, commercial sector, agriculture, spacecraft, aircraft, smart grid, etc. The major challenge is to keep up the environment very smart and ease-of the accessibility of the services as well. To provide a better solution, sensor technology can be implemented in the required and critical sectors. Sensors are very tiny, affordable, and intelligent in sensing the data from various regions even humans cannot reach out to. Raw sensory data provided by the sensors are transformed into useful information for further processing. Remote sensor systems are intended to gather data by methods for a huge number of vitalities restricted to battery sensor hubs. Hence, it is imperative to limit the

Intelligent Sensor Node-Based Systems: Applications in Engineering and Science, Anamika Ahirwar, Piyush Kumar Shukla, Prashant Kumar Shukla, and Ruby Bhatt (Eds.)

vitality devoured by every sensor to broaden the system's life. The objective of this work is to structure an intelligent sensor node-based system that gathers data. However, such data could reasonably be expected to be processed wisely. Three important phases in intelligent sensor technology are equipping the sensors in the critical/required areas, sensing the data, and transforming the sensory data into useful information for the end-user. Most of the critical regions like nuclear power plants, the military sector, unmanned aerial vehicles (UAV), health monitoring, and agriculture require intelligent technology in this fast-growing environment. Hence, this technology is considered a key to every sector.

1.1 INTRODUCTION

In recent years, wherever the transmission and reception of a huge amount of data are involved from disparate sources, human intervention is minimized, and sensory technology is employed to leverage the existing framework. This technology has been marked as an innovator for the recent advancement in a wide variety of areas. With the development of low power, less cost, small-sized, and sensors intelligent in nature, manifold fields deploy these intelligent systems in real-world applications. Currently, several open issues are there in intelligent sensor node (ISN)-based systems are briefed below:

1. **Platform of the Sensor Node:** The primary issue is the platform of the sensor node. It is about the most efficient way to schedule and upgrade a kind of less costly hub than the two well-known phrases of the Berkeley Motes and iPaq-based sensor node.
2. **Energy Efficiency:** The other key issue is energy efficiency. It is a critical problem of sensor networks. Sensor networks, such as processing units, antennas, sensors, and actuators, have various sources of energy consumption. Strategies at the hub level and network-level procedures are used for various executive strategies for efficient resources.
3. **Space and Time Issues:** The next main issue is space and time issues. Time synchronization is a basic part of the sensor network's base. Virtually all forms of combination of sensor information and composed incitation involve coordinated physical time to think

about physical world occasions. The prerequisites for clock precision and accuracy are regularly more relevant in sensor systems than in conventionally acceptable frameworks. The space issues include the limitation of the hub and issues with sensor inclusion. They are both important in supporting sensor system administrations and applications.

4. **Protocol of Sensor Systems:** Then the other important issue is the protocol of sensor systems. The convention stack contains the physical layer, the layer of the information interface, the layer of the arrangement, the transport layer, and the layer of the application. The physical layer refers to the specifications of the fundamental but effective tweak, transfer, and acceptance strategies. Since the earth is boisterous and sensor hubs can be portable, the convention for medium access control (MAC) must be power-mindful and ready to restrict crashes with communications from neighbors. The device layer manages to direct the information provided by the vehicle.
5. **Collaborative Signal Processing:** The last challenge is collaborative signal processing. To gather and process information to produce helpful data, the nodes in the sensor orchestration must cooperate. The extent of data exchange between hubs and how hubs integrate the data from different hubs are significant specialized issues. Similarly, the trade-offs between the better implementation of the system and asset limitations in the handling of mutual signs and data must be considered. Overall, it makes the system smarter.

In engineering, such as industrial automation, safety surveillance, unmanned aerial vehicle (UAV), health monitoring, military purpose, commercial use, home automation, etc., this intelligent sensor node-based system reached a very large step, which in turn provides a smart environment [2]. In all of these systems, intelligent sensors play a major role.

The major objectives of the ISN-based system include:

i. A plan for wireless intelligent sensor and actuator network (WISAN) and development of an economical setup of instrumentation for the accomplishments of structural patient care monitoring.

ii. The procedure is strongly applied to permit the arrangement of new WSNs to have the option to respond and to adjust highlights relying upon the environmental conditions, keeping up quality of service (QoS), and extending the lifetime of the network.
iii. Improvement of a structural patient care technique, which is reasonable for self-sufficient basic healthcare monitoring.

These intelligent sensors collect sensory information from multiple networks and various locations where a human cannot reach the target and notify the status of the process to the end-user. This intelligent technology is implemented in various sectors wherever continuous monitoring is mandatory and inevitable.

For instance, in the healthcare sector, continuous attention is required for the patient by monitoring and measuring the parameters such as blood pressure, temperature, heartbeat rate, etc. In this case, this system supports collecting the values and sends the information to the clinical setting for further analysis/diagnosis. This facilitates healthcare professionals to analyze the patient status and attend to the cases in case of an emergency immediately. Likewise, the involvement of this technology in every realm utilizes the system effectively with minimum affordable cost.

1.2 WHAT IS THE INTELLIGENT SENSOR NODE (ISN)?

Intelligent sensors are the sensors that are capable of sensing or gathering raw data from various network environments created on a user-defined procedure and communicate with every other node connected in a network as useful information. These sensors are the key portion of any kind of Intelligent Sensor-based System. Hence a node that is equipped with several intelligent sensors is called an intelligent sensor node (ISN).

1.3 HISTORY OF ISN

For a long time, sensor nodes have been present and commonly used in various applications, from earthquake measurements to armed supply warfare. In the late 1990s, when the Smart Dust Project and the NASA Sensor Webs Project are still active, the newest creation of small sensor nodes appeared. One of the primary objectives of the smart-dust project is the development of autonomous sensing and communication within a cubic millimeter of space. Although it didn't receive a great ending, it led to research for many projects till now. The remarkable projects developed via smart dust [3] act as a base for establishing major research centers. A physical sensor node is also referred to as a mote or pod by researchers. Physical sensor nodes can expand their capacity related to Moore's Law. The chip footprint contains additional difficult and low-powered micro-controllers. In this way, for a similar hub impression, more silicon ability can be stuffed into it. In recent days, sensor nodes are used to serve long possible wireless ranges (dozens of km) with low energy consumption.

Present-day sensor nodes are second-generation commercial products from the fourth phase. The advancement in computing and communication in the late 1990s and early 2000s resulted in the new invention of sensor node technology. Developing a sensor network signifies a major improvement over the traditional sensor. Reasonably priced small-sized sensors based on micro-electromechanical systems (MEMS) are highly attractive and widely used. The present energizing advances in the IC plan make it conceivable to scale down sensor gadgets with estimating, calculation, and correspondence abilities. Although every sensor has restricted capacities, the arrangement of an enormous wireless sensor network (WSN) achieves various muddled capacities in a wide range of applications, for example, in natural, horticultural, or mechanical systems, observation, and control. On implementations, sensor hubs regularly have no earlier information on their position, and in this manner, a confinement instrument is frequently a prerequisite.

1.4 RELATED TERMINOLOGIES

Current advances in remote and electronic advances have empowered a wide scope of utilization of ISN-based systems in military detection, traffic observation, target following, disorder observation, social insurance check, etc. Here this system depicts such sort of propels in ISN and their applications in different fields such as healthcare, military, industrial/commercial, healthcare, agriculture, and so on.

1.4.1 LITERATURE SURVEY

- Authors Y. E. E. Ahmed et al. in the paper "3D Virtual Biomimetic Network: A Topology for Resilient Intelligent Wireless Sensor Networks," stated that Remote information correspondence has a key job in any kind of refreshed data innovation-based venture. Versatility, flexibility, adaptability, and simplicity to introduce are the principal highlights of remote information procurement frameworks. While the constrained lifetime, accessibility, disappointment hazard, and unwavering quality are as yet critical difficulties. Roused by the cobwebs as an appealing common correspondence area, this work proposes utilizing the bug-catching network's geography to structure dependable and versatile, clever remote sensor systems to be executed for remote information securing, intending to give promising answers for such issues. The chapter gives a few definitions to develop ideas and pertinent documentation, for example, the virtual string, virtual string thickness, virtual string thickness, sensor hubs, and base station (BS) sending, information briefest way, information way versatility, and dependability, from cobwebs perspective [11]. At that point, it portrays the huge examination difficulties to be tended to for material 3D Virtual bug catching networks. Additionally, it represents the customization approach of 3D virtual bug-catching networks of remote information procurement for administration, security, asset arrangement, and dynamics.
- Authors S. Devendra et al. in the paper "Design and Development of WINGSNET (wireless intelligent GPS-based sensor network) System for Monitoring Air Pollution and Radiation Based on Wi-Fi and WiMAX Communication Network," stated that The Wi-Fi and WiMAX communication networks are based on the 'WINGSNET' (wireless intelligent GPS-based sensor network) system for the detection of air pollution (NO_2, NO, O_3, CO, CO_2, SO_2, PM10, PM25) and radiation. The WINGSNET system is designed to track air pollution and nano-sensors of radiation spread over the topographic area. The sensor units/nodes (SU) are competent and adequately dedicated to the processing of the sensor data and the determination of the status, as usual, warning or alarm status and reporting of the ID, location, status, date/time

of the Sensor to the portable/fixed reporting station (RS). The RS sends status data to a centralized monitoring station (CMS) via the BS. CMS will progressively display the sensor status for each field and generate the Status-Report/Service-Report as required. The sensors can be fixed or mobile with the assistance of the Wi-Fi Cell/WiMAX Cell/GPS System and the positions of the stationary sensors are pre-characterized as a zone/sub-zone/X-Y grid while the mobile/airborne sensor locations are resolved. The Sensor-Units will be solar-fueled or battery-based [12] with low power consumption. The communication/acceptance of data between RS and CMS is specified by a Messaging System/Format. As a 'MAHAWIRG' system (MOBILE AD HOC ACCESS WIRELESS GPS-BASED), consisting of MSs (mobile stations) and BS, which differ from region to region, a subset of WINGSNET can be configured. With 'Broadband Wireless Technology' and Bi-directional (Symmetric) Communication Links, the proposed WINGSNET/MAHAWIRG system is based on Wi-Fi/WiMAX. For instance, voice, data, messages, graphs, images, video, and multimedia are communicated and received by a few applications on the device.

- The authors Z. Weng et al. in the paper "Design of node controller for wireless monitoring system of central air conditioner," stated that in a savvy house, one of the essential and simple hardware is central air conditioning. To make the cooling framework function in an efficient and vitality-saving state, the shrewd control framework for focal cooling has been studied depending on WSN. The middle section of the focal cooling control framework [13] is the hub regulator. It understands the terminal gear control and the variety of signals, and the remote correspondence with the hubs of the steering. The equipment circuit and programming structure of the note regulator is proposed in this chapter, and a model is presented. The remote continuous cooling framework observation and control are recognized, and the model's exploratory results indicate that the framework is running steadily.
- The authors H. Sharma et al. in the paper "An Efficient Solar Energy Harvesting System for Wireless Sensor Nodes," stated that the WSNs in genius systems, shrewd stopping, and keen urban areas are the critical structure squares of the new internet of

things (IoT) base. The WSN hubs suffer the ill effects of a major plan imperative that their battery vitality is reduced and can run depending on the duty pattern of operation for only a few days. They propose another answer to this structural problem in this chapter, using sun-powered photovoltaic vitality [14]. Here, for battery-powered battery-based WSN hubs, they suggest a deeply efficient and one-of-a-kind sun-oriented vitality recovery system. The improved solar energy harvesting wireless sensor network (SEH-WSN) hubs could operate for an infinite device lifetime in an ideal world (in years). They suggest in this chapter a novel and powerful sun-oriented fueled battery-accusing arrangement for WSN hubs of the most severe force point following (MPPT). The examination center is about to extend the efficiency of the general gathering system, which relies on Solar Panel Efficiency, DC-DC converter productivity regulated by MPPT also battery-powered battery skills. A few models have been developed for the sun-powered vitality reaper system and iterative recreation has been performed in MATLAB/SIMULINK for sun-based managed DC-DC converters with MPPT to achieve ideal performance. It is seen from the recreation outcomes that our planned vitality reaping system based on sunlight has 96% efficiency (η sys).

- The authors T. P. Lambrou et al. in the paper "A Low-Cost Sensor Network for Real-Time Monitoring and Contamination Detection in Drinking Water Distribution Systems," stated that a simple and all-encompassing way to resolve the issue of observing water quality for drinking water appropriation frameworks as well as buyer destinations. Their methodology relies on the creation of minimal-effort sensor hubs for continuous and in-pipe inspection and on-the-fly evaluation of water quality. A few electrochemical and optical in-pipe sensors consist of the main sensor hub and emphasis is placed on minimal effort, lightweight use, and solid long-term service. Such usage is suitable for large-scale organizations that empower the sensor organization approach to provide spatiotemporally rich data to water users, water organizations, and specialists. To identify minimum effort sensors that can accurately screen a few limits, which can be used to create the water quality, large writing and statistical surveys are carried out. Because of the selected boundaries, together with a few microsystems for simple sign molding, planning, logging, and far-off introduction of details, a sensor display is created [15]. Finally, to assess the water tainting threat, calculations are generated to combine multi-sensor estimates at the neighborhood level on the web. On deliberate contamination occasions of various convergences of Escherichia coli microbes and overwhelming metals, studies are carried out to determine and authorize these estimates (arsenic). Test results show that at truly low fixations, this modest structure is appropriate for separating these high-effect pollutants. The results show that this structure fulfills the perfect early warning system of the online, in-pipe, low sending activity cost, and great discovery accuracy steps.

- The authors M. Sajid et al. "Remote monitoring of environment using multi-sensor wireless node installed on the quadcopter drone," stated that in the area of atomic force facilities, environmental observation and radiation discovery is a questionable errand and there is a possible danger of adding radiation to individuals and untamed life. Ecological limits such as humidity, temperature, radiation, and so on are screened remotely. The sensors were mounted on a quad-copter drone that was remotely operated. Besides, GPS was implemented to assess the area and height of the automaton for the unique boundaries [16]. As a multi-sensor independent remote hub, an Arduino Yun-based circuit interface was structured and mounted on the automaton. The Yun Board's Wi-Fi was used to transfer all the details to the BS. The Android-based application was built to talk to the hub and continuously view the information along with logging in to the gadget. Also, through printed gadgets techniques, a moisture and temperature sensor was manufactured and additionally installed on the automaton to replace business sensors. The analysis work aimed to assemble a structure that could be used in the future to analyze the area of atomic force plants and to research the development of ecological boundaries in that district. The authors Anitha Kumari and Sudha Sadasivam discussed applying classical and quantum cryptographic techniques for the internet of medical things [25].
- The authors D. Pašalić et al. in the paper "ZigBee-based data transmission and monitoring wireless smart sensor network

integrated with the Internet," stated that the paper considers and portrays possible outcomes and approaches to structure and actualize a ZigBee-based knowledge transmission and observation of remote shrewd sensor arrangements integrated with the Internet. The use and reconciliation of different equipment components and programming dialects are needed to upgrade a financially savvy structure. In the chapter, possible outcomes and main sections of genius sensor hubs are considered and portrayed. The mix of distant, savvy sensor arrangements and their Internet sensor hubs are expressed in more subtleties [17]. A cheap vitality-saving ZigBee highlight point remote shrewd sensor device is designed and executed to screen sensor information in a vitality-productive manner using XBee modules and various sheets are proposed and described. The setup of XBee modules used for correspondence between shrewd sensor hubs is also illustrated. Common-sense use of keen sensor hubs and facilitator hubs powered by a battery is represented in subtleties. Additionally, data estimation with such sensor hubs is seen. The chapter appears after the results of remote information estimation, information movement, and control with such remote shrewd sensor-based on ZigBee organizing employing Network advancements.

- The authors J. Liang et al. in the paper "A Distributed Intelligent Hungarian Algorithm for Workload Balance in Sensor-Cloud Systems Based on Urban Fog Computing," stated that it is possible to create new structures with the aid of fog computing, urban computing, and insight to enhance the urban condition and quality of human life. Urban fog figuring-dependent sensor-cloud frameworks (SCS-UFC) are modern savvy organizing frameworks that consolidate a cloud process with WSNs as well as fog hubs to provide enormous-reaching capabilities such as detecting, measuring, and data capability. Since the sensor hubs only have limited bandwidth in WSNs, their data cannot be easily communicated to the cloud level. Consequently, to move the data from WSNs to the cloud level, fog hubs with more grounded bandwidth are conveyed [18]. In any event, different outstanding tasks (i.e., information measurements) can be disrupted by extraordinary fog hubs: typically, the fog hubs with heavier remaining tasks at hand mean longer transmission deferment and more vitality usage. If a fog hub depletes its vitality, it will bite the dust and then make the machine stop working. In this way, the remaining load of all fog hubs must be balanced to minimize the postponement of transmission and the use of the sensors for vitality. Nevertheless, tending to the problem is testing because each fog hub only knows its neighbors' nearby results, and it is, therefore, difficult without anyone else to obtain a worldwide advancement result. A disseminated canny calculation based on the Hungarian technique is suggested in this chapter. First of all, each fog hub collects data related to its neighboring fog hubs that are located within its go transmission. Another inherited measurement is intended at that stage to find an approximate streamlining arrangement. Finally, each fog hub decides whether it can advance parts of its remaining load to other fog hubs to change the outstanding tasks at hand for all fog hubs. The results of recreation indicate that our estimate will achieve shorter deferral and less use of vitality than current works.

- The authors A. Kumar et al. in the paper "An Energy-Efficient Smart Comfort Sensing System Based on the IEEE 1451 Standard for Green Buildings" stated that Comfort is a significant point of view in building automation, and the continuous calculation of solace is famously intertwined. They have built up a remote, savvy comfort-detecting system in this chapter. The important limits were considered in the planning of the prevailing estimate of solace frameworks, such as ease, power consumption, unwavering efficiency, and framework cost. Based on the IEEE 1451 standard, the correspondence module, sensor hub, and sink hub were generated to achieve the objective plan goals. For the enhancement of the sensor cluster, electrochemical and semiconductor sensors were considered and the after-effects of the two developments were studied. Using the ATMega88 microcontroller [19], the sensor and sink hubs were performed. The graphical UI in C# was developed using Microsoft Visual Studio 2013. Following the sign preparation circuit, the sensors were calibrated to ensure that normal sensor accuracy was achieved. This chapter provides answers to problems that occurred in the writing, point-by-point plan.

- The authors L. Ma et al. in the paper "Distributed Event-Based Set-Membership Filtering for a Class of Nonlinear Systems with Sensor Saturations Over Sensor Networks," stated that A class of discrete time-differing framework with an occasion-based correspondence system over sensor systems is examined for the conveyed set-participation separating problem. The feasible structure depends on the limited nonlinearity of the section, obscure but limited clamors, and sensor immersions. Each informative detection hub sends the data to its neighbors only when there is an abuse of certain set-off conditions. Adequate conditions are calculated by methods for several recursive framework imbalances for the existence of the ideal suitable occasion-based channel that is fit for limiting the framework state in some evaluation-focused ellipsoidal districts. Two additional enhancement issues are found within the built-up hypothetical structure: one is to look for the negligible ellipsoids (in the sense of network follow) for the best separating exhibition, and the other is to amplify the setting off-limit to lessen the setting of recurrence with good sifting execution [20]. To take care of the streamlining problems, a numerically attractive confusion calculation is used. Finally, to illustrate the feasibility and validity of the proposed algorithm, an illustrative model has been implemented.
- The authors A. Javed et al. in the paper "Design and Implementation of a Cloud-Enabled Random Neural Network-based Decentralized Smart Controller with Intelligent Sensor Nodes for HVAC," stated that building energy management systems (BEMSs) screen and control the ventilation and air conditioning (HVAC) of heating structures, because of different other structural mechanisms and services. WSNs have become the fundamental piece of BEMS at the underlying level of use or last when retrofitting is needed to overhaul more evolved systems. Nonetheless, WSN-empowered BEMS have a few problems monitoring data, regulators, actuators, insight, and force utilization of remote segments (which may be battery controlled). For implanting insight into the sensor hubs, the remote sensor hubs have minimal preparation force and memory. A random neural network (RNN)—brings together a keen regulator in this chapter concerning an IoT stage integrated with cloud preparation to prepare the RNN that was performed

and tried in a domain chamber. However, the IoT stage is basic and not restricted to having a few temperature estimation sensors, stickiness, channel air originating from the HVAC conduit, and PIR. There are three key components to the smart RNN regulator: (i) the base station; (ii) sensor hubs; and (iii) the cloud with inserted information for different errands on each segment. With cloud planning for the preparation of the RNN, this IoT stage is integrated. In the center of the sensors, an RNN-based habitat estimator is implanted, which assesses the number of inhabitants within the room and sends this data to the BS [21]. With RNN models, the BS is inserted to monitor the HVAC based on warming and cooling set points. In comparison to simple principle-based regulators, the Nature Chamber HVAC expends 27.12% less vitality with an intelligent regulator. This proposed half-breed inhabitancy estimation calculation that consolidates the RNN-based inhabitancy estimator with the entry sensor hub reduces the inhabitancy estimation time (furnished with PIR and an attractive reed switch). The results suggest that the crossover RNN inhabitancy estimator accuracy is 88%.

- The authors T. W. Foster et al. in the paper "A Web-Based Office Climate Control System Using Wireless Sensors," stated that as a rule, daily heating, ventilating, and air conditioning (HVAC) frameworks achieve the optimal degree of control employing basic ON-OFF control strategies, which can also result in a high waste of vitality. Clever automatic HVAC regulators that base their activities/choices on sensor information are a probable response to this issue. An office environment monitoring and control system is designed and executed in this document. An office environment monitoring and control system is designed and executed in this document. The system consists of various hubs for remote sensors and a control hub [22]. The sensor hubs provide crucial information for the sensor to determine inhabitancy, and the control hub conducts the calculation, which concludes whether the sensor information allows cooling or warming. As a regulator, this system can fill in and can be integrated into HVAC frameworks in savvy structures. The created control calculation performed on the control hub has been shown to boost the productivity of vitality by up to 39% over standard ON-OFF regulators for HVAC frameworks.

- The authors S. Hu et al. in the paper "Integrated Determination of Network Origin – Destination Trip Matrix and Heterogeneous Sensor Selection and Location Strategy," stated that a two-phase improvement model for the dynamic (camera-based tag recognition) and aloof (vehicle indicator) sensor data to test the origin-destination (O-D) trip grid and the heterogeneous sensor arrangement procedure in an optimized way for the organization of vehicular traffic. To decide the ideal methodology for the organization of the sensor, the main stage describes the heterogeneous determination of the sensor and the problem of the region to increase the traffic data available for the issue of O-D lattice estimation as far as the quantity of the two sensor types and their establishment areas are selected. The traffic data includes the observed link stream, path of way, and way of inclusion data. This traffic data is used in the subsequent stage to create the O-D system grid that limits the error between the observed and evaluated traffic streams (interface, O-D, as well as way). Correspondingly, two lattice estimation models of the O-D device are proposed where the link-based model consolidates the stream preservation rule between O-D and connection streams and uses the connection core occurrence network, and the way-based model assumes a certain connection-way frequency grid. The system's O-D structure and connection stream gauges are intended to be calculated. Findings from computational simulations propose that the way-based model beats the link-based model in the estimation of system O-D frameworks. Besides, the general responsibilities of the mixtures of the two sensor types to the problem of system O-D grid estimation are examined [23]. The outcomes of the study are primary consequences for heterogeneous sensor determination and area systems.
- The authors G. Loubet et al. in the paper "Implementation of a Battery-Free Wireless Sensor for Cyber-Physical Systems Dedicated to Structural Health Monitoring Applications," stated that for the automated physical systems dedicated to the auxiliary well-being observing applications in cruel circumstances, the proposed device tends to the concept of a remotely controlled and

sans battery remote sensor. The suggested material engineering relies on a sophisticated remote sensor system of detection hubs and conveyor hubs [24]. To detect the real environment, the detecting hubs are used. They are sans battery and, employing a far-field remote force transmission system, remotely controlled by a committed radiofrequency source.

The information collected by the detecting hubs is sent to the distribution hubs that, among other things, link the physical world through the Internet to the computerized world. Using a Lora WAN uplink remote correspondence and temperature and relative humidity sensor, a detecting hub model was developed, and the tests were performed to represent it. The exploratory results show that by using only the remote force transmission downlink, the periodicity of estimate and correspondence can be controlled remotely. In this chapter, the authors present this complete use of a remote-fueled and battery-free remote detection center – not yet integrated or scaled-down intended for the actualization of total physical-digital structures and depending on the remote data and force movement at the same time. Finally, an analysis of the almost equivalent use of the hubs without battery detection for the digital-physical structures has been completed.

- The author S. Bosse, in the paper "Distributed Agent-Based Computing in Material-Embedded Sensor Network Systems with the Agent-on-Chip Architecture," stated that new information handling and correspondence designs are required in distributed material-implanted frameworks such as sensor systems coordinated in sensorial materials. In the face of the hub, sensor, device, information preparation, and correspondence disappointments, unwavering quality, and power of the entire heterogeneous condition must be provided, particularly regarding limited assistance of material-implanted frameworks in the wake of assembly. In this chapter, multi-agent frameworks with state-based portable specialists are used as a rule comprising a solitary microchip to register in untrustworthy work such as hub systems, providing a novel plan approach for solid transmitted and equivalent knowledge preparing for inserted frameworks with static assets [26]. To prepare the specialist behavior for multi-agent structures completely implementable on microchip-level supporting agent on-chip (AoC) preparation models, a propelled elevated-level amalgamation system is used. Using a reconfigurable pipeline conveying process engineering performed with restricted state machines and register-move justification, the operator actions, affiliation, and flexibility are fully organized on the microchip. The nature of the operator preparation is characterized by Petri Net symbolic handling. Any level of specialist variance and algorithmic determination is achieved by a reconfiguration feature of the operator handling system. For an action-based specialist behavior programming language, the operator behavior, affiliation, and portability highlights are shown and indicated. The relation and correspondence of the operator are generated by a simple tuple-space database performed at the hub level and signals giving far away correspondence and collaboration at the hub level.

1.5 COMPONENTS OF ISN

The following are the basic elements of ISN as shown in Figure 1.1:

- Sensing element;
- Device micro-processing unit;
- Communication.

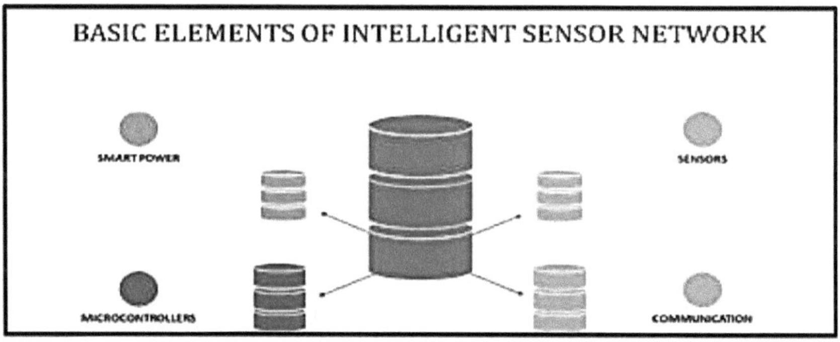

FIGURE 1.1 Basic elements of an intelligent sensor network.

The major hardware components of any intelligent node-based system are stated below. These components are suitable for various diversified applications and add-on sensors can be additionally equipped depending on the application:

1. **Power Supply Unit:** Though the sensors are tiny, any electronic device requires a power supply. This unit is powered by batteries.
2. **Transceiver Unit:** It associates the node with the framework.
3. **Sensing Unit:** This unit consists of two additional subunits: (i) sensors; and (ii) analog-to-digital converter to detect the raw data that is likely to be transformed for further prediction/analysis into useful information.
4. **Processing Unit:** It is the central management unit of the entire sensing task.
5. **Memory Unit:** Any processing unit requires storage memory. This unit is used to store both raw data and useful information as well.

1.6 BASIC WORKING OF INTELLIGENT SENSOR NODE (ISN)

As far as the current development is concerned, sensor networking technology increases rapidly due to its suitability and minimum cost. The intelligent sensor-based system uses numerous sensor nodes implemented over a specific region depending upon the application and user requirements. Intelligent sensors are of three types, namely: (i) sensors with the capacity to predict; (ii) sensors with learning capability; and (iii) sensors with innovative capability. Based on the user requirement and suitability to the environment any kind of sensor can be chosen based on the application/requirement. As these sensors are exceptionally small and moderate it tends to be utilized successfully in any setup. Sensor nodes start collecting information from various sources once they are deployed in the required region. This supports the user to monitor the status of the requirement such as heat, temperature, light, object tracking, speed, direction, etc., based on the application. Any intelligent system has the basic components as the same, apart from its more number of additional sensors equipped depending upon the user-created procedure or application. Hence the design of the ISN-based system is modular. The nodes in the system communicate effectively with every other node via wired or wireless. The sensors connected to any environment are said to be very intelligent, as it has the capability of sensing and notifying the information through the sensors. Every sensor functions in four functioning modes: (a) transmission; (b) reception; (c) idle listening; and (d) sleep. The sensors consist of a tiny operating system to drive any kind of hardware device and perform various tasks. This goes about as a middle layer between the equipment sensors and the client/administrator. Sometimes collision may occur due to the simultaneous transmission of values from two or more nodes. A simple structure of a sensor-based system is shown in Figure 1.2.

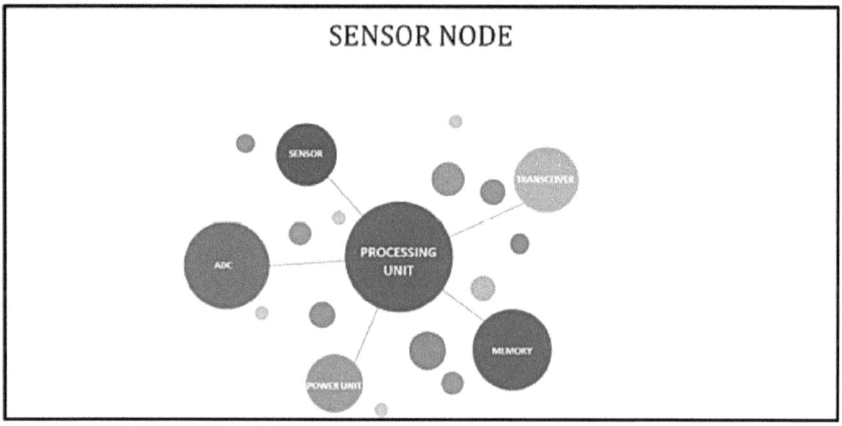

FIGURE 1.2 A simple structure of an intelligent sensor-based system.

1.7 CHARACTERISTICS OF INTELLIGENT SENSOR NODE (ISN)-BASED SYSTEM

The key physical characteristics of the sensor node used to assess the performance of the ISN-based system are stated in Table 1.1.

TABLE 1.1 Key Characteristics of Intelligent Senor Node-based System

Characteristics	Description
Fault acceptance	Every node connected to the network is prone to failure. The level of fault acceptance plays a very big role in any intelligent sensor node-based system. Fault acceptance is the ability to withstand the sensor network performance without any interruption in case of any sensor node miscarriages.
Movement of the nodes	Depending upon the application, the movement of the nodes will be present anywhere in the system. This helps the sensor-based system to achieve efficient communication.

TABLE 1.1 *(Continued)*

Characteristics	Description
Modular network topology	Sensor nodes in the network are built in compliance with standard topology. In dynamic topology, the system should have the capacity to operate.
Communication miscarriages	If the transfer of data to any other node is not feasible in the case of any node in the system, the information should be notified to the base station or gateway node without any further suspension.
Heterogeneity of nodes	The intelligent sensor nodes equipped in any system are of different types and they all work together in a synchronized manner.
Scalability	Intelligent sensor networks are equipped with a huge number of sensor nodes even in hundreds or thousands. Therefore, the intelligent system is supposed to be extremely scalable.
Self-driven/ autonomous	The intelligent system must work autonomously without any central point to control.
Ability to reprogram/ reconfigure	The system should have the ability to reprogram or reconfigure in case of any changes that occur in the network.
Use of sensors	The intelligent sensors connected to the network should be consumed at the maximum with high performance and minimum energy consumption.

1.8 TYPES OF SENSOR NETWORKS

Depending upon the properties of the node, sensor networks are classified into two types. They are:

- homogeneous sensor networks; and
- heterogeneous sensor networks.

1.8.1 HOMOGENEOUS SENSOR NETWORKS

The sensor nodes present in this type of network have similar properties in terms of communication, storage, processing, energy level, and trustworthiness. If the entire sensor set has similar properties, then this type of network is homogeneous and called a homogeneous network. An example of a homogeneous sensor network is shown in Figure 1.3.

FIGURE 1.3 Homogeneous sensor network.

1.8.2 HETEROGENEOUS SENSOR NETWORKS

The sensor nodes present in this type of network have dissimilar properties in terms of communication, storage, processing, energy level, and trustworthiness. If the entire sensor set has dissimilar properties, then this type of network is heterogeneous and so-called heterogeneous network. A sample network is shown in Figure 1.4.

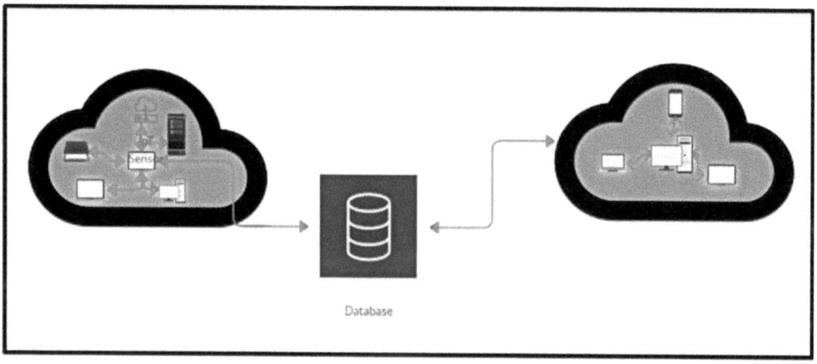

FIGURE 1.4 Heterogeneous sensor network.

1.9 SENSING SCHEMES

Data collection is one of the major functionalities of any sensor technology. Sensors collect information from several environments by applying any one of the appropriate sensing schemes. The sensing scheme is another important parameter in communication.

The sensing schemes are of different types as follows:

- Compressive sensing scheme (classical);
- The adaptive intelligent compressive sensing scheme;
- Reshuffling clustered compressive sensing.

1.9.1 COMPRESSIVE SENSING SCHEME

This type of sensing scheme is the classical sensing scheme. Here the collected data or signal is compressed at a specific level only. The sparsity of the signal is static in this scheme. Hence the fixed transformation is done with the help of the prior information that is less suitable for a real-time environment.

1.9.2 ADAPTIVE COMPRESSIVE SENSING SCHEME

This type of sensing scheme is the extension of the classical compressive sensing scheme that is said to be adaptive and intelligent because the collected sensory data can be reconstructed under the temporal and spatial domain [4]. Here in this scheme, it is a must to set a few criteria to attain the successful recovery of the signals.

1.9.3 RESHUFFLING CLUSTERED COMPRESSIVE SENSING SCHEME

By combining compressive sensing with cluster procedure, the system can achieve higher efficiency and support wider networks as well. Based on the features – cluster head performance and cost of the pre-treatment, the raw data can be reconstructed with very minimum samples [9]. Since it uses reshuffling pre-treatment, the collected sensory data can be of significantly better quality.

1.10 EFFECTS OF ISN

The deployment of an intelligent sensor-node-based system creates a great impact on a wide variety of fields in recent years. These intelligent systems are well-equipped in a suitable environment that is capable of monitoring, communication, and processing. The managing sector may be any of the sectors like government, civil, industrial, and commercial. This intelligent technology can be deployed wherever transmission and reception between nodes are mandatory. Most of the areas such as the military, healthcare, environment, home automation, commercial sector, etc., produced much more impact over the years. By implementing this intelligent sensor technology in recent thrust areas, human intervention can be reduced so that is prone to dangerous situations. In some critical regions like smoke or fire detection, human intervention is highly perilous wherein this sensor technology creates a great impact and saves a huge number of human lives. This technology was first implemented in the military sector, where confidential information was gathered from various places that too unknown to the critical region. Intelligent sensors play a major role in collecting this information and reaching the BS or gateway where human involvement is very dangerous. One another critical region where human intervention is very difficult is the nuclear power plants. This intelligent sensor technology plays a key portion of it that is used to collect sensory information where that humans cannot do. Therefore, advanced state-of-the-art, promising technologies such as ISN-based systems are widely implemented in both engineering and science. Most of the sectors show active performance because of this sensor technology. On the whole, this system is used extensively in every sector due to its high performance and effectiveness at an affordable cost.

1.11 ISN MAJOR DOMAINS

The major domains of ISNs are numerous. Generally, the holistic framework observes, tracks, and controls the environment. In a common application, the Intelligent Sensor-Node-based framework is sent to a zone where it is intended to gather information through its sensor hubs. The following are the few application domains of ISNs:

- Environmental observation and forecast;
- Catastrophe management;
- Structure control for well-being;
- Accommodation monitoring;
- Monitoring of the field;
- Eco-friendly tracking;
- Greenhouse tracking;

- Landslide detection;
- Industrial control;
- System health control;
- Monitoring for aquatic/wastewater;
- Monitoring of landfill ground well level and pump counter;
- Farming domain marine tracking;
- Intelligent ventilation;
- Regulation of lighting;
- Extensive traffic stream control in huge metropolitan areas;
- Supply-chain control in state-of-the-art manufacturing plants.

1.12 APPLICATION SECTOR

There are several intelligent sensors available such as temperature sensors, humidity sensors, pH sensors, multimedia sensors, etc. As a result, an ISN-based system is deployed in various applications in critical sectors like environment monitoring, agriculture monitoring, industrial monitoring, military operations, health monitoring, home automation, etc. The subsequent section discusses in detail the functioning.

1.12.1 MILITARY

Primarily these intelligent sensor networks are intended for surveillance and military purpose. This has been considered a brilliant tool for military and air force applications that comprises monitoring various constraints, gathering information, surveillance of the war field, and attack detection, etc. Due to their competencies in real-time transmission, this ISN-based system performs a key role in the military sector. These intelligent networks have numerous benefits such as robustness, fault tolerance, and low-cost deployment.

The main objective in the military sector is to detect, classify, and track the attacker at the border [10]. Also, confidential information needs to be collected from the target where sensor technology plays a significant role. A battlefield deploying this ISN-based system can detect the presence, count, location, track, and identify the attacker, and most importantly sense the critical information. Considering these criteria, the system should be developed, which in turn helps to overcome the challenges like field noise, field variation, and background signals in the military sector. A sample scenario is shown in Figures 1.5–1.7.

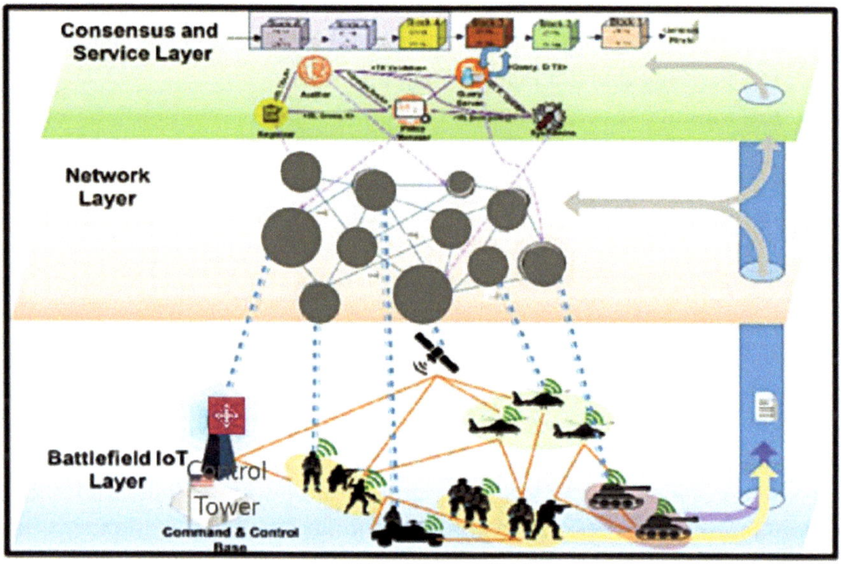

FIGURE 1.5 Sensors in the military sector.

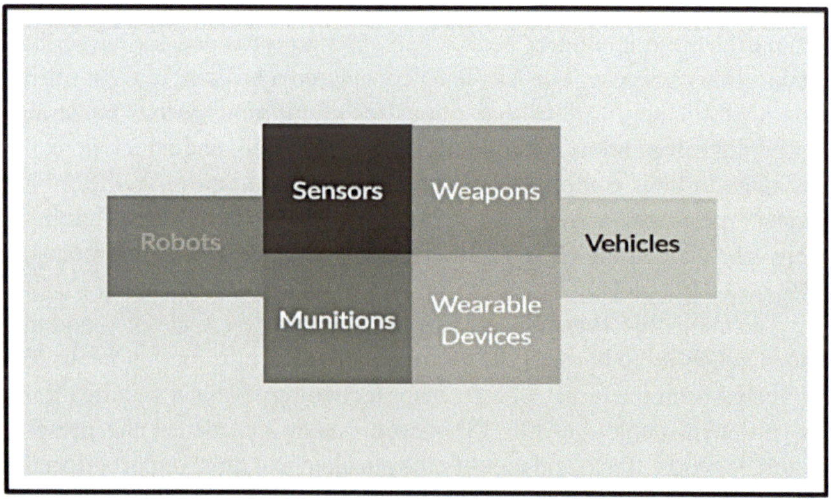

FIGURE 1.6 Devices with sensors on the battlefield.

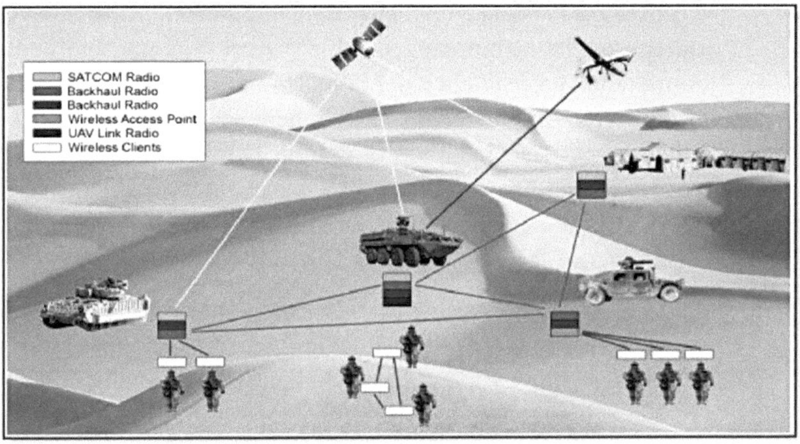

FIGURE 1.7 Sensor technology in the critical region.

1.12.2 HEALTHCARE

This ISN-based system plays a key role in the healthcare sector. This system can be well utilized in healthcare for:

1. **Health Monitoring:** By implementing this system in the healthcare sector, both healthcare personnel and patients can be benefited. This supports the practitioner or caretaker to monitor the patient's health in a clinical setting and home setting as well. It is essential to frequently verify the patient's vital parameters like heartbeat rate, blood pressure, body temperature, etc. To verify such parameters, intelligent sensors and location tags are placed in the monitoring system.
2. **Intelligent Nursing Homes:** Another major issue is to taking care of elderly people. This wireless technology with intelligent sensors addresses this challenge by providing affordable healthcare services and supporting them to live autonomously. The intelligent sensors will sense all the information that helps to detect any unusual patterns or behavior virtually.
3. **Tele-Care:** This support system uses information and communication technologies to perform clinical work. This allows faraway as well as virtual medical evaluations. Also, this healthcare scheme minimizes global costs.

4. **Wireless Body Area Networks:** By using several intelligent wearable sensors in the human body, this system supports continuously monitors the patient's health at the hospital or home. This system is well suited for emergency cases where healthcare personnel can gather the patient's health information. This also provides support in several healthcare services like cancer identification, medical data access, memory boost, asthma identification, and monitoring blood glucose [8].

In reality, the healthcare sector experiences numerous challenges like shoot-up costs, growing incidence of medical errors, inadequate staffing, an aging population, etc. Despite all the challenges, medical practitioners use these modern-day technologies to offer better services [7]. Pervasive healthcare can minimize durable costs and provide betterment in the quality of service (QoS). Intelligent sensor networks provide the ubiquitous healthcare system with effective solutions. By implementing this system in the healthcare sector, both healthcare personnel and patient can be benefited. This supports the practitioner or caretaker to monitor the patient's health in a clinical setting and home setting as well. Figures 1.8 and 1.9 posturize the usage of intelligent sensors in the healthcare sector.

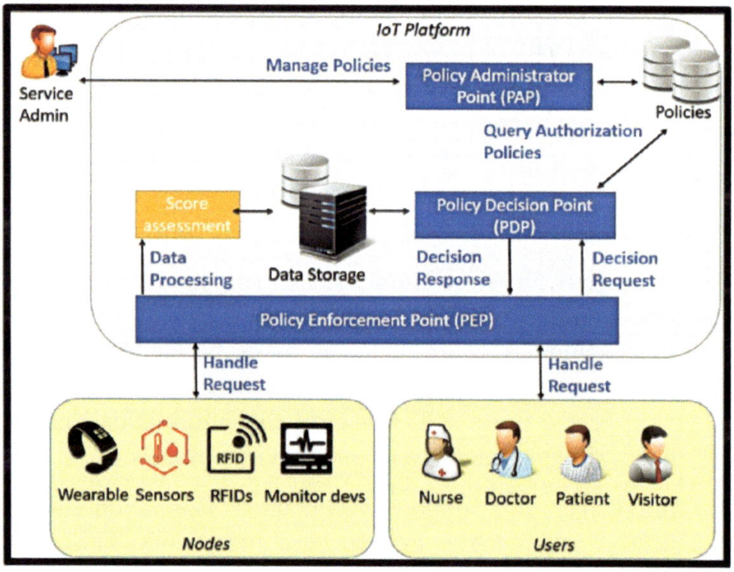

FIGURE 1.8 Intelligent sensor technology in the healthcare sector.

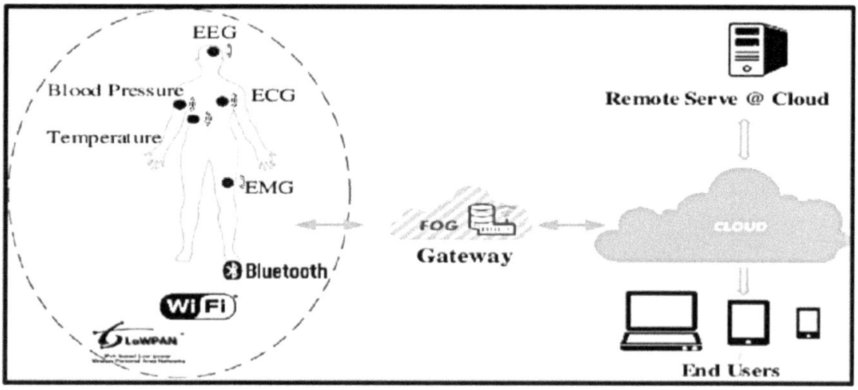

FIGURE 1.9 Sensor technology for observing health parameters.

1.12.3 ENVIRONMENT

An Environmental crisis is one of the major concerns in almost every country for the past few years. Most importantly, water and air are the two main constituents of the environment. Due to the establishment of numerous factories, the quality of the water and the air is worsening day by day. To face these challenges, it is a must to monitor the production progression and environmental bounds of the factory sector. It is a cumbersome process and hazardous as well. Here comes the role of the sensor technology that addresses this issue. The factories with the help of these intelligent sensors acquire environmental data. Continuous monitoring of the environment improves the industrial process [6]. This technology also offers better services for watershed monitoring systems, health monitoring in rivers, energy managing solutions, and atmospheric discharges. Besides, it serves as a better solution for air pollution monitoring, forest fires monitoring, greenhouse monitoring, and landslide detection. A simple picture is presented in Figure 1.10 to show the intelligent sensor technology in the environmental sector.

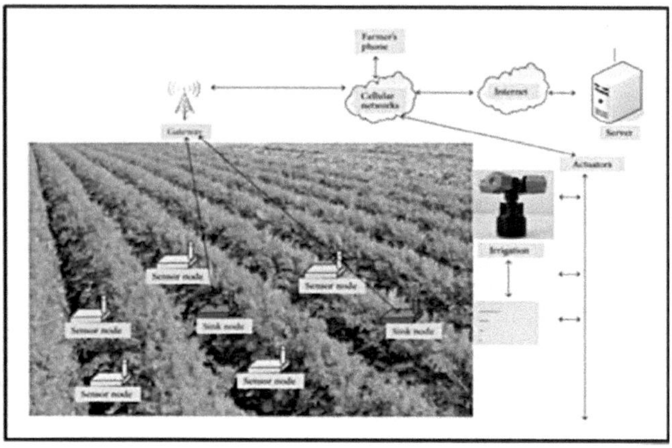

FIGURE 1.10 Sensor technology in the environmental sector.

1.12.4 HOME AUTOMATION

In our daily life, the parameters like light, temperature, humidity, etc., are the major portion where human communicates and tries to control them materially. Home automation can be set up via communication and information technology through the internet that supports monitoring these parameters [5]. Several intelligent sensors will be placed in various appliances according to the user's requirement, which in turn notifies the user by providing sensory information. This system provides people relaxation, and security as well as provision for energy saving by monitoring daily consumption. A simple scenario for home automation is shown in Figure 1.11.

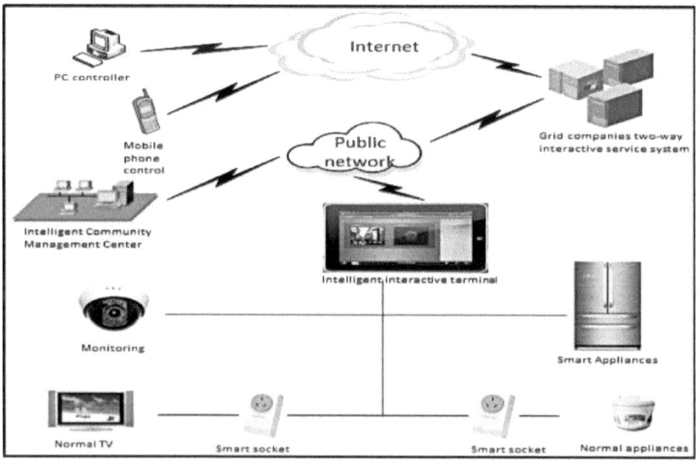

FIGURE 1.11 Home automation scenario.

1.12.5 COMMERCIAL USE

The evolution of sensor technology has improved a lot by the way a product is used before. Innovations are steadily increasing day by day for the betterment of the services. Now a day's sensor technology is instrumental in controlling various areas such as factories, offices, homes, vehicles, cities, and the overall ambiance, specifically in almost all areas. For instance, a few areas or buildings that are prone to earthquake can equip earthquake-related sensor to protect the structural change and also helps to save human lives. Self-detector can also be placed inside the buildings to identify the structural fault.

Likewise, any natural disaster like a tsunami can also be notified by implanting this sensor technology. Such a kind of notification is very essential in saving a huge number of human lives. In bank sectors, this sensor technology is especially used for security surveillance. By implementing this technology, both clients and the banks get benefited profoundly. Better services can be provided for the clients with the latest updates, and most importantly, waiting time will be reduced. In the automotive industry, autonomous functioning enhances using this sensor technology. On the whole, automated services are everywhere across the world to reduce the intervention of humans. The usage of sensor technology for commercial purposes is shown in Figures 1.12 and 1.13.

1.12.6 USE CASES OR CASE STUDIES

Some of the use cases that are implemented successfully in real-time are listed below:

- Sensor nodes demonstrating high-temperature characteristics and wide temperature;
- Human footstep sound classification;
- Pedestrian counting/number of pedestrian traveling;
- Design and realization of ISN with application in energy awareness using WSNs.

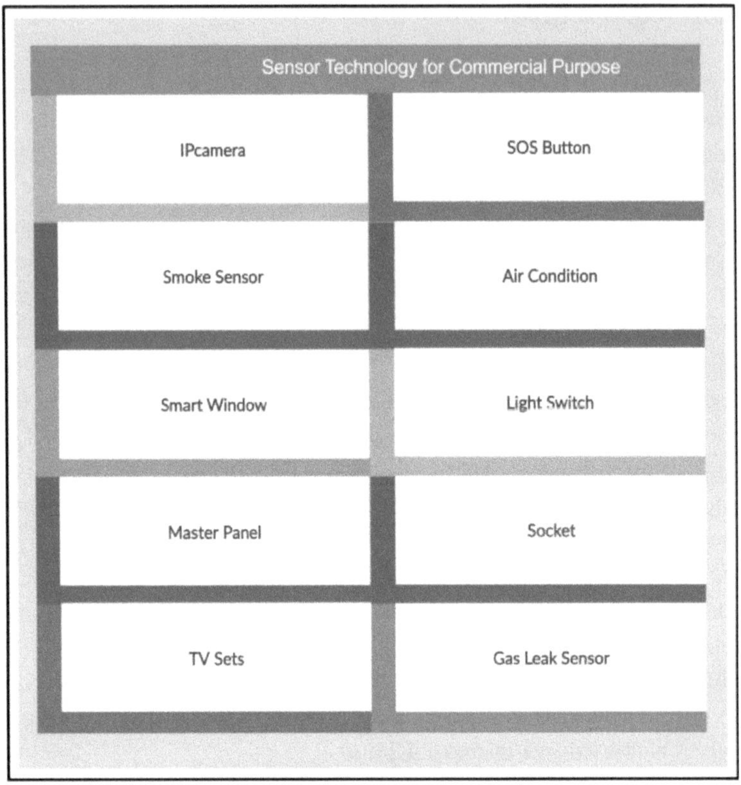

FIGURE 1.12 Sensor technology for commercial purposes.

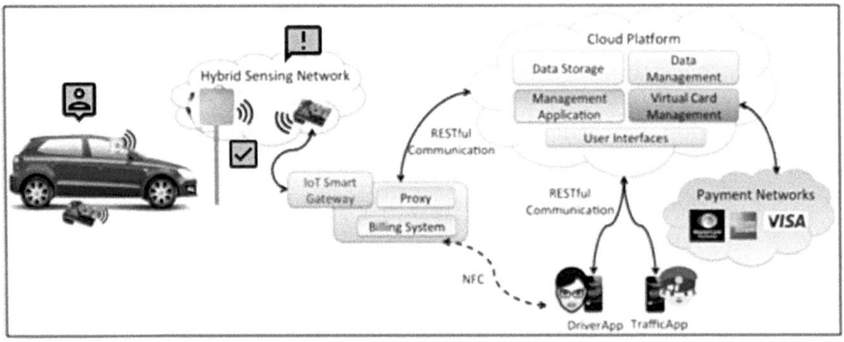

FIGURE 1.13 Gathering data from the vehicle (car).

1.12.6.1 SENSOR NODES DEMONSTRATING HIGH-TEMPERATURE CHARACTERISTICS AND WIDE TEMPERATURE

Temperature has a major function in various research fields and daily lifestyles. It is an essential size index within the fields like chemical, medical, food field, and equipped fields. There are many forms of temperature sensors to be had of which the maximum are stressed or lively sensors. This requires a physical connection between the sensors and signal transmission machine with strength deliver as a vital unit. Though it satisfies a few necessities, it fails to satisfy some special necessities. There is a short lifespan and operating range of stressed sensors. In addition to sensors such as platinum resistors, thermocouples, optics, surface acoustic wave (SAW), and LC (inductance and capacitance) resonance, there are several degree temperature strategies (Figure 1.14).

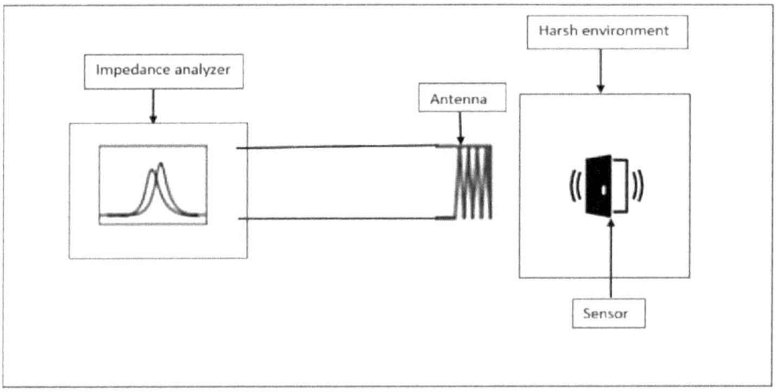

FIGURE 1.14 Proposed wireless coupling system for temperature measurement.

There are precise benefits of wireless passive LC resonant temperature sensors. Especially in aggressive industrial and scientific settings, it is highly suitable for short-distance electricity transmission to meet requirements such as revolving components, unfavorable control, chemical corrosion, sealed setting, and other special occasions for high temperatures. High-temperature co-fired ceramics HTCC [1] tapes are manufactured for use in lamination and sintering methods as a substrate. It also manufactures a Wi-Fi passive temperature sensor that incorporates a planar spiral inductor and a parallel plate capacitor. A wireless coupling method is used to calculate the efficiency of the temperature sensor.

1.12.6.2 HUMAN FOOTSTEP SOUND CLASSIFICATION

Sensor nodes present in a WSN for surveillance applications are required to be small, power-efficient, and inexpensive with enormous computational abilities. Along with this, an appropriate records processing scheme in the sensor node allows for decreasing the energy dissipation of the transceiver through the compression of information to be communicated. In this application, a simulation-based examination of a human's footstep sound category in a normal environment has been tried by employing consuming very simple time-domain features. The Wi-Fi sensor nodes in a WSN are designed according to the required target application. Sensor nodes for safety surveillance software for monitoring the presence of people in a restricted sector fall beneath the category of passive supervision. These sensors are normally small-sized static sensor nodes and are to be deployed in massive numbers with continuous tracking. These resource-limited sensor nodes have to be less expensive and strength efficient (a few μW) for serving the purpose (Figure 1.15).

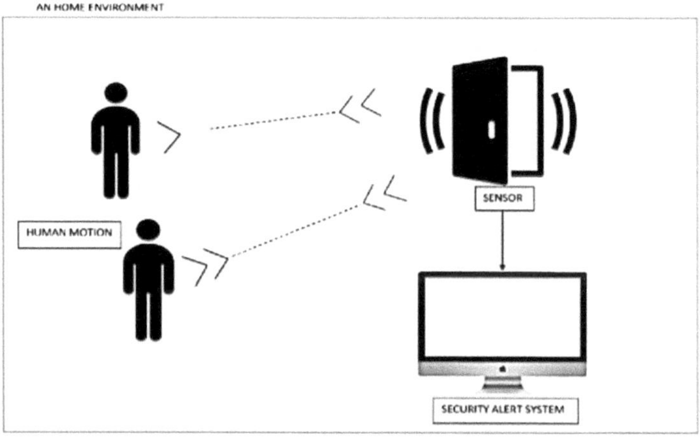

FIGURE 1.15 Scenario for human footstep sound classification.

An integrated device layout for a weak sensor node focused on a surveillance sector is briefed here. The node is supposed to identify the human presence in restricted areas by evaluating the generated acoustic audio indicators using low complexity time-domain capabilities accompanied by using analog domain neuromorphic implementation. The proposed set of rules is in comparison with standard strategies to verify in terms of accuracy. The simple hardware components are as follows: DS function vector generator, slice generator, cut up generator, splice deal with a generator, weights memory, IF neutron array.

1.12.6.3 PEDESTRIAN COUNTING/NUMBER OF PEDESTRIANS TRAVELING

A shrewd sensor hub operated by the IoT has been exceptionally designed and developed for keen metropolitan territorial applications. The sensor hubs are manufactured to check the number of walkers, and their path of excursion along the edge of some simple limits. The Fresnel focal point field of view (FoV) should be economically accessible in passive infrared (PIR) sensors. They have been specifically tuned to screen canine, feline, and so on developments of only individuals and no other home creatures. The surrounding boundaries include temperature, mugginess, pressure, carbon dioxide (CO_2), and a total volatile organic portion (TVOC). Through the Long-Range Wide Area Network (Lora WAN) discussion system, the tested measurements are passed to the web worker. To achieve a precision of 95% for the individual on foot, a keen calculation has been advanced.

The use of a video, the computerized camera is one of the most obviously used methods of human tallying. It equally has an ample assortment of utilizations inside the observation of capacity territories. A warm computerized camera is some other mainstream procedure for human recognition. It is additionally utilized for vehicle recognition, looking for casualties around evening time, spotting seething flames inward a divider, and identifying overheating electric wiring. The warm camera's abilities by methods for distinguishing the infrared force transmitted by utilizing an item which is known as a warmth signature and building up a photo electronically dependent on data around the temperature contrasts. PIRs are other basic innovations for human recognition. These low-fee movement sensors answer when an infrared transmitting subject (individuals or creatures) goes through its Freight on Value (FoV). PIR sensors had been utilized employing numerous analysts for human discovery, restriction, and walk speed assessment at indoor reconnaissance.

The concept here is a 3 m wide trail and restricts the length of the location to 1 m from the road level to take off the home-grown creature at any cost. To widen the hub for tallying individuals on foot and finding the direction of travel, three sets of PIR sensors are used. The sensor hub is set up at a distance of 4 m from the electrical shaft. Instead of using this technique when locating the PIR sensors, the inclusion positions of the sensors are now not hidden. The blunders may, therefore, be reduced. Base and center PIR sensor sets are run to decrease affectability, but to limit the blunder, the apex pair operates at maximum affectability. Individuals in the wheel seat also fall under the recognition district of Sensors. Although the proposed structure has countless benefits, there are a couple of challenges as well. This structure does not differentiate between individuals and creatures with an extra stature of 1 m. Kids snoozing on the coach are not counted because they hold 1 m of freedom from the street level (Figure 1.16).

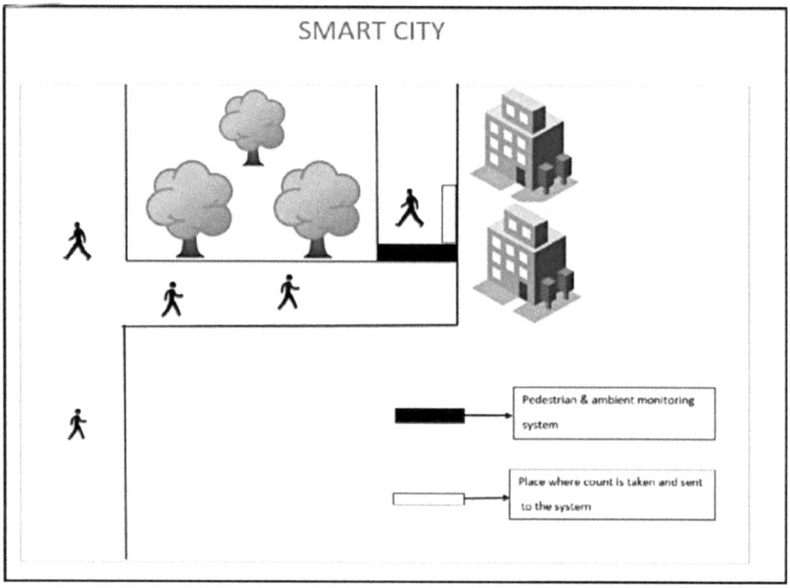

FIGURE 1.16 Pictorial representation of the direction of travel and environmental monitoring system for a smart city.

1.12.6.4 DESIGN AND REALIZATION OF ISN WITH APPLICATION IN ENERGY AWARENESS USING WIRELESS SENSOR NETWORKS (WSNS)

In the various commercialized business automation techniques and numerous other real-life packages, the era of wireless sensors plays a crucial role. It is especially suitable for mission-critical environment packages in which it is difficult and/or almost impossible to deploy other network infrastructure, including battlefields, hazardous chemical plants, and high thermal environments. It is not always unusual to see that most critical security and safety systems often rely entirely on sensor-based packages. Sensors that are small in size and inexpensive in price have the potential to be deployed in several applications. Inherently, all sensor networks have some sort of sensing mechanism, either by a time-driven technique or by an occasionally triggering approach, to collect facts from an intended physical environment. A sensor transmits the sensed data to the desired target spot or sinks via these techniques. The features of the WSN are peculiar to traditional networks.

Such basic features are also taken into account to resolve problems and challenges related to community coverage, management of runtime topologies, distribution of nodes, administration of nodes, power efficiency/consumption node mobility, community deployment, and application areas/environment (Figure 1.17).

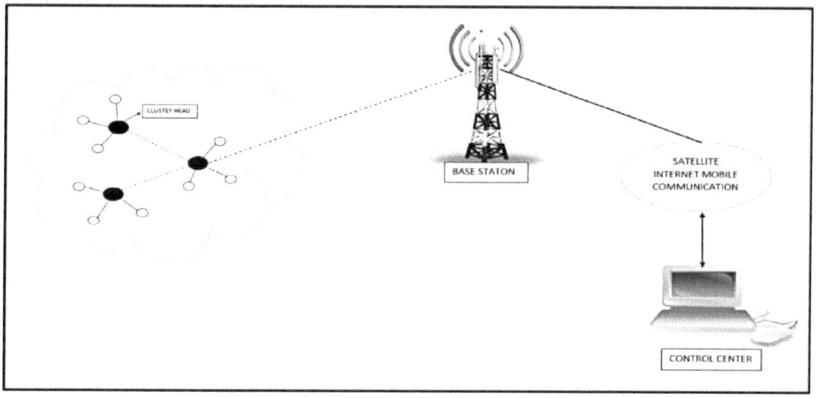

FIGURE 1.17 A modern WSN.

1.13 CONCLUSIONS AND FUTURE WORK

The Intelligent sensor-node-based system implemented both in engineering and science has reached a very big step in this technological world. It provides better services in every public or private sector. Since this technology involves very tiny, intelligent, and affordable sensors, it can be equipped in any small network to an enterprise network, which in turn provides higher efficiency and faster process with minimum cost.

Most of the critical regions require this kind of technology wherever human interference is prone to danger. This also can save a large number of human lives as well. In sensor frameworks, existing sensor progress is inclined towards increased unpredictability. Organized electronic innovation enables PC handling that once demanded enormous and refined sign-preparing systems to be reduced to a microelectronic chip, for example, intelligent transduction sensors, signal enhancement, sifting, and other preparation of a solitary substrate. The more remarkable adaptability, and lower manufacturing costs associated with cutting-edge, coordinated electronic innovation. From the point of view of the end-user, even with its increased utility and inner intricacy, the sensor system currently seems more straightforward.

To be used in applications such as submerged acoustic sensor frameworks, detection of cyber-physical frameworks, time-basic applications, intellectual detection and range of executives, and security and safety of executives, future developments in the sensor node-based system must produce incredible and economically intelligent gadgets. All possibilities of further development in ISN applications can be examined in this field of investigation.

KEYWORDS

- **commercial purpose**
- **critical region**
- **healthcare**
- **intelligent environment**
- **medium access control**
- **military**
- **sensing scheme**
- **sensors**

REFERENCES

1. Tinggui, C., Gongfa, L., Jianjun, Y., & Honghai, L., (2016). Wireless smart sensor networks, systems, trends, and the applications in engineering. Hindawi Publishing Corporation. *Journal of Sensors,* 1526153.

2. Carlos-Mancilla, M., López-Mellado, E., & Mario, S., (2016). Wireless sensor networks formation: Approaches and techniques. Hindawi Publishing Corporation. *Journal of Sensors*, 2081902.
3. Mohammed, S. B., Raoudha, S., Yessine, H. K., & Mohamed, A., (2020). Wireless sensor network design methodologies: A survey. Hindawi Publishing Corporation. *Journal of Sensors*, 2081902, 9592836.
4. Jian, C., Jie, J., Yansha, D., Xingwei, W., & Abdol-Hamid, A., (2018). *Adaptive Compressive Sensing and Data Recovery for Periodical Monitoring Wireless Sensor Networks*. MDPI.
5. Sandeep, P., Heye, Z., Md Eshrat, E. A., Hemant, G., Subhas, C. M., Yuan-Ting, Z., & Wanqing, W., (2017). A novel secure iot-based smart home automation system using a wireless sensor network. *MDPI Sensors, 17*, 1–19.
6. Muhammad, S. J., Muhammad, A. J., Anam, M., Ahsan, I., Abdullah, A., & Usman, M., (2015). Smart environment monitoring system by employing wireless sensor networks on vehicles for pollution free smart cities. Humanitarian technology: Science, systems and global impact 2015. *Hum. Tech., 107*, 480–484.
7. Surajudeen, A. A., & Kurubaran, G., (2019). *m-Health in Public Health Practice: A Constellation of Current Evidence*. Chapter 11, B978-0-12-816948-3. 00011-8.
8. Matthew, N. O. S., Kelechi, G. E., & Sarhan, M. M., (2018). Wireless sensor networks for healthcare. *Journal of Scientific and Engineering Research, 5*(7), 210–213.
9. Lu, Z., Baishan, C., Yuanyuan, L., & Zhizhang (David), C., (2015). Data gathering in wireless sensor networks based on reshuffling cluster compressed sensing. Hindawi Publishing Corporation. *Journal of Sensors*, 260913.
10. Baseline, P. S. R., Pradeep, M., & Gajendran, E., (2016). Military applications of wireless sensor network system. *A Multidisciplinary Journal of Scientific Research & Education, 2*(12).
11. Ahmed, Y. E. E., Adjallah, K. H., & Amin, M. B. M., (2019). 3D virtual biomimetic network: A topology for resilient intelligent wireless sensor networks. In: *2019 10 IEEE International Conference on Intelligent Data Acquisition and Advanced Computing Systems: Technology and Applications (IDAACS)* (pp. 1002–1006). Metz, France.
12. Devendra, S., Verma, K., & Barhai, P. K., (2014). Design and development of WINGSNET (wireless intelligent GPS-based sensor network) system for monitoring air pollution and radiation based on Wi-Fi & Wi-MAX communication network. In: *2014 IEEE 11 International Conference on Mobile Ad Hoc and Sensor Systems* (pp. 507–508). Philadelphia, PA. doi: 10.1109/MASS.2014.118.
13. Weng, Z., Fang, J., Weng, Z., Cheng, Y., & Zhang, G., (2017). Design of node controller for wireless monitoring system of central air conditioner. In: *2017 IEEE International Conference on Cybernetics and Intelligent Systems (CIS) and IEEE Conference on Robotics, Automation and Mechatronics (RAM)* (pp. 361–364). Ningbo. doi: 10.1109/ICCIS.2017.8274802.
14. Sharma, H., Haque, A., & Jaffery, Z. A., (2018). An efficient solar energy harvesting system for wireless sensor nodes. In: *2018 2 IEEE International Conference on Power Electronics, Intelligent Control and Energy Systems (ICPEICES)* (pp. 461–464). Delhi, India. doi: 10.1109/ICPEICES.2018.8897434.

15. Lambrou, T. P., Anastasiou, C. C., Panayiotou, C. G., & Polycarpou, M. M., (2014). A low-cost sensor network for real-time monitoring and contamination detection in drinking water distribution systems. In: *IEEE Sensors Journal* (Vol. 14, No. 8, pp. 2765–2772). doi: 10.1109/JSEN.2014.2316414.
16. Sajid, M., Yang, Y. J., Kim, G. B., & Choi, K. H., (2016). Remote monitoring of environment using multi-sensor wireless node installed on quadcopter drone. In: *2016 IEEE International Symposium on Robotics and Intelligent Sensors (IRIS)* (pp. 213–216). Tokyo. doi: 10.1109/IRIS.2016.8066093.
17. Pašalić, D., Bundalo, D., Bundalo, Z., & Cvijić, B., (2015). ZigBee-based data transmission and monitoring wireless smart sensor network integrated with the internet. In: *2015 4 Mediterranean Conference on Embedded Computing (MECO)* (pp. 240–243). Budva doi: 10.1109/MECO.2015.7181913.
18. Liang, J., Long, Y., Mei, Y., Wang, T., & Jin, Q. (2019). A distributed intelligent Hungarian algorithm for workload balance in sensor-cloud systems based on urban fog computing. *IEEE Access*, *7*, 77649–77658, doi: 10.1109/ACCESS.2019.2922322.
19. Kumar, A., & Hancke, G. P., (2014). An energy-efficient smart comfort sensing system based on the IEEE 1451 standard for green buildings. In: *IEEE Sensors Journal* (Vol. 14, No. 12, pp. 4245–4252). doi: 10.1109/JSEN.2014.2356651.
20. Ma, L., Wang, Z., Lam, H., & Kyriakoulis, N., (2017). Distributed event-based set-membership filtering for a class of nonlinear systems with sensor saturations over sensor networks. In: *IEEE Transactions on Cybernetics* (Vol. 47, No. 11, pp. 3772–3783). doi: 10.1109/TCYB.2016.2582081.
21. Javed, A., Larijani, H., Ahmadinia, A., Emmanuel, R., Mannion, M., & Gibson, D., (2017). Design and implementation of a cloud-enabled random neural network-based decentralized smart controller with intelligent sensor nodes for HVAC. In: *IEEE Internet of Things Journal* (Vol. 4, No. 2, pp. 393–403). doi: 10.1109/JIOT.2016.2627403.
22. Foster, T. W., Bhatt, D. V., Hancke, G. P., & Silva, B., (2016). A web-based office climate control system using wireless sensors. In: *IEEE Sensors Journal* (Vol. 16, No. 15, pp. 6104–6113). doi: 10.1109/JSEN.2016.2574896.
23. Hu, S., Peeta, S., & Liou, H., (2016). Integrated determination of network origin–destination trip matrix and heterogeneous sensor selection and location strategy. In: *IEEE Transactions on Intelligent Transportation Systems* (Vol. 17, No. 1, pp. 195–205). doi: 10.1109/TITS.2015.2473691.
24. Loubet, G., Takacs, A., & Dragomirescu, D., (2019). Implementation of a battery-free wireless sensor for cyber-physical systems dedicated to structural health monitoring applications. In: *IEEE Access* (Vol. 7, pp. 24679–24690). doi: 10.1109/ACCESS.2019.2900161.
25. Anitha, K. K., & Sudha, S. G., (2018). A comparative analysis of classical cryptography vs. quantum-safe cryptography. In: *Medical Big Data and Internet of Medical Things: Advances, Challenges, and Applications* (pp. 299–325). CRC Press, Taylor & Francis.
26. Bosse, S., (2014). Distributed agent-based computing in material-embedded sensor network systems with the agent-on-chip architecture. In: *IEEE Sensors Journal* (Vol. 14, No. 7, pp. 2159–2170). doi: 10.1109/JSEN.2014.2301938.

CHAPTER 2

IoT-Based Smart Chair for Healthcare Supporting System

POOJA GUPTA,[1] SUNITA VARMA,[2] NEERAJ ARYA,[1] and UPENDRA SINGH[1]

[1]*Assistant Professor, IT Department, Shri G. S. Institute of Technology and Science, Indore, Madhya Pradesh, India*

[2]*Professor and Head of IT Department, Shri G. S. Institute of Technology and Science, Indore, Madhya Pradesh, India*

ABSTRACT

Incorrect posture is one of the significant problems in our day-to-day life. The current observation is that due to long sitting on a chair, people suffer from some back issues, which makes their posture incorrect. Thus we need a solution for detecting the posture of a person sitting on a chair. Hence, we are making this in real-time by connecting it to the chair with an electronic circuit. The system uses Arduino UNO, two sets of Ultrasonic sensors, Bluetooth module HC-06 to send messages to the person using the chair, and a buzzer/speaker to indicate bad posture. We used Bluetooth GSM's Android mobile app to connect Bluetooth module HC-06 with our mobile. Bluetooth module sends the text message 'take the rest' every two-hour. The message is sent to the user's mobile phone. Since we are developing a healthcare project, we have also added the features of checking temperature and pulse rate while sitting on the chair. The idea behind adding this feature is that it is very beneficial for the person suffering from heart rate problems. Then, the person can use the band attached to the chair to check the pulse rate. Another advantage is that this product is very beneficial for

Intelligent Sensor Node-Based Systems: Applications in Engineering and Science, Anamika Ahirwar, Piyush Kumar Shukla, Prashant Kumar Shukla, and Ruby Bhatt (Eds.)

old age people. If the temperature goes high, then a message is sent to the doctor. We use Lm-35 and a pulse rate sensor to check temperature and pulse rate.

2.1 INTRODUCTION

In India, the production of any organization is dependent on its employees. Employees work hard towards their organizations to fulfill their objectives. As one of the world's most populous countries, India has many office professions that involve lengthy and constant hours of sitting. Employees spend significant amounts of time leaning over their computers or laptops, which has a substantial negative impact on their back and neck muscles.

In a survey of 200 people, it is configured that 90% have back pain because of continuously sitting in their workplace. Not only back pain is the issue, but a repetition of such poor practice of sitting with the wrong posture has changed their body structure from a straight to a curved back. Another problem that emerged was sitting constantly with a straight or correct posture is also harmful to the back, so rest is needed after every two hours (Figure 2.1).

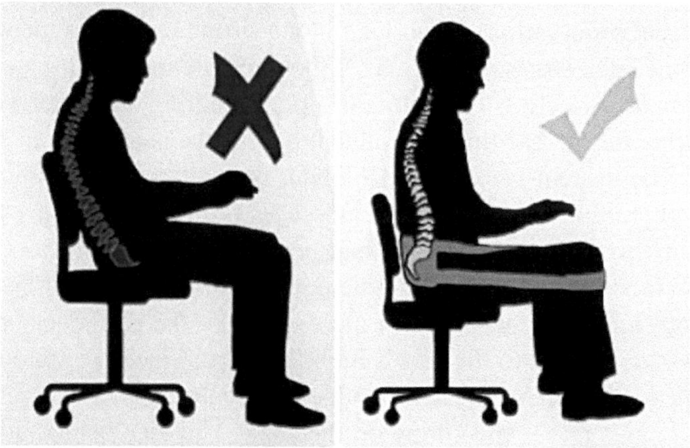

FIGURE 2.1 Position of chair sitting.

To eradicate all back-related problems and prevent the wrong posture, we need to develop a system that ensures the correct posture and gives notification to take rest after every 2 hours of sitting. As a result, we have created a methodology for the chair in this work, so we designate it as a

Smart chair. In conjunction with sensing a person's body temperature and heartbeat, this chair assists in improving their posture.

2.1.1 COMPONENTS USED

Here in the proposed model, we are employing the following sensors in order to detect a specific type of object. The temperature sensor, oxygen, and heart rate sensor are made here using the following components for a specific task:

1. **Temperature Sensor (LM 35):** High-temperature measuring thermometer that is contact based. The LM35 is a measure temperature equipment that generates analog power that is proportional to the temperature. Allows output from Centigrade to be used (Celsius). The system does not rely on any outside measurement cycles to function (Figure 2.2).

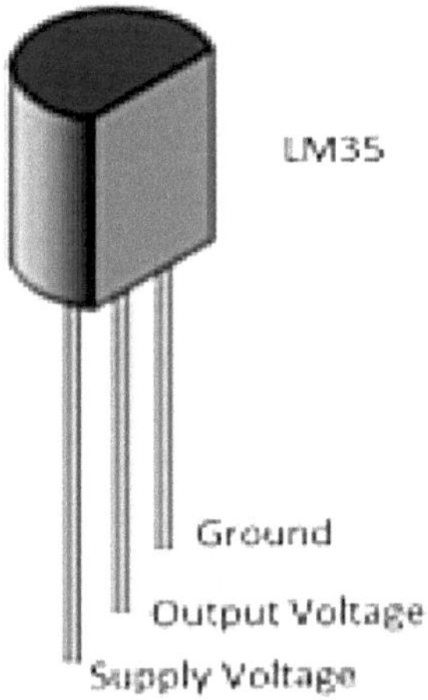

FIGURE 2.2 LM35 sensor.

2. **Heart Rate Sensor (Pulse Sensor):** Although it is employed in a simple and direct manner, the location of this sensor is critical for the accurate measurement of pulse rate. The sensors need to be covered with a conductive material like hot glue, vinyl tape, or other moving material. With the Heart emblem on the front, the positive side of things was sensed. That's the side of the body that

comes into contact with the skin. A little round hole is seen on the front of the device, whereby the LED shines (Figure 2.3).

FIGURE 2.3 Pulse sensor.

3. **Ultrasonic Sensor:** This is one of the sensors used for measuring distance and for object sensing [1]. As a result, we have implemented this component in the smart chair system. This sensor works on the simple formula, i.e., distance = speed × time. This sensor gives out an ultrasonic wave, and if any object comes between the wave and the sensor, the wave is reflected into the sensor, as well as the ultrasonic receiver picks up on this wave and begins operating, i.e., identifying the object (Figure 2.4).

FIGURE 2.4 Ultrasonic sensor.

4. **Bluetooth HC-06 Module:** Bluetooth module is an II-class-based Bluetooth server developed for serial communication of wireless. When it is associated with a Bluetooth device, including a PC or laptop, a phone, or a tablet, the user will be able to see and use its features. Each and every piece of information received through serial setup is delivered and over the air without delay. Whenever the modules acquired data transmission, they transmitted it to the host computer over a serial interface in the same manner that it was received. A simple user control method does not necessitate the employment of a Bluetooth code module that is particular to the user. The HC-06 can function with a voltage supply ranging from 3.6 VDC to 6 VDC; although, the RXD pin frequency is 3.3 V and cannot sustain a supply voltage of more than 5 V (Figure 2.5).

FIGURE 2.5 HC-06 module.

5. **Arduino UNO Board:** Arduino is mostly known for its open-source microcontroller source board, which is founded on the Semiconductor ATmega328P microcontroller. Arduino is also known for its open-source Arduino shield. Input signals connections as well as an analog/output (I/O) connector are provided on the board, which can be interrupted by a variety of additional associations for the advancement (shields) and other rings. The

board features 14 digital I/O connectors (six of which have PWM output), and 6 I/O pins, and is bundled with the Arduino IDE (integrated development environment), which connects through a B-type connector. It also has a USB port. Despite the fact that it can operate at voltages ranging between 7 to 20 volts, the device can be recharged using a USB connection or an additional battery with a power of 9 volts or higher (Figure 2.6).

FIGURE 2.6 Arduino UNO.

6. **Buzzer/Speaker:** A buzzer or speaker is an audio signature instrument; electrical, electromechanical, or piezoelectric actuators are all possible (piezo for short). Warning systems, timers, and user input verification, such as a mouse click or key, are all examples of common usage for buzzers and speakers (Figure 2.7).
7. **Software Used:** For the execution of all of the above mentioned components, we are applying the Arduino Development Environment. Using this Arduino IDE, you may connect every pin to any other pin on your Arduino board. Eagle can also be used for the circuit and printed circuit board design.

FIGURE 2.7 Speaker.

2.2 LITERATURE REVIEW

The authors of this chapter have developed a traffic control system that is based on radio frequency identification (RFID). When an emergency occurs, this system is in charge of managing and regulating traffic lights on the road. They have developed a scenario in which the ambulance is tracked using GPS, and the location of the ambulance is communicated to traffic signal management, which subsequently turns on the blue light on the traffic signal when the ambulance passes by, informing onlookers there's an emergency [2].

Here a system is designed for the HPV driver, and the HPV driver will transfer a signal to the system. This signal will then be responded to based on priority. The signal will cause the green light to illuminate, allowing the car to pass through without interfering with traffic flow. This system is analyzed on the SUMO, and 50% of the time is saved in various situations [3].

In this study, the authors offer a system based around Android and the cloud, and it also makes use of GSM technology to make it more affordable. The system makes use of a GSM module, MQTT, a microcontroller, traffic signals, as well as an Android mobile phone for communication. Using the proposed technology, the lights can be controlled in line with the current situation [4]. The authors have indicated that the proposed model can be used to create a true-time system for the transmission of data on traffic situations and the observation of these variables. In this manner, the use of cameras and sensor network technology is employed, and the clearing time for all lanes through which the ambulance will pass is checked. Priority for ambulance vehicles is obtained through the RFID. The given system is constructed using a graphical user interface (GUI), making it simple to use [5]. The proposed traffic monitoring system in this chapter has given a very intelligent method to deal with traffic problems. The authors stated that the fixed interval of the traffic light needs to be intelligent and sense the traffic in the present situation. It is necessary for the signal lights to manage the timings according to the amount of traffic present; in addition, emergency vehicles such as ambulances should be identified, and the movement of ambulances should be given precedence [6]. Using the sensor, the authors of this research have created a sanitizer device that also includes a temperature sensor as part of the overall system. The COVID-19 disease is being monitored by this system, which is aimed to take preventative steps. TMP36 temperature sensors are employed in this system for temperature calculation, while ultrasonic sensors are used in the sanitizing machine [7]. The authors have provided an intelligent ambulance system for traffic control in this section. It is possible to use this model for both health monitoring and traffic control. When they took the fever, they decided to take it to the emergency room. For traffic signaling, radio frequency transmitters and receivers are utilized. This signaling system is utilized to determine the current traffic situation, and the ambulance is directed to the least congested route [8]. The authors of this research conducted measurements of temperature, heart rate, and other parameters. All of this information is then transmitted through GSM to any mobile phone with a GSM connection. Vehicle tracking is accomplished through the use of GPS and GSM, with programming carried out in assembly language [9]. The operation of this system is as follows: a microcontroller is fixed for monitoring values, and when any limit is exceeded, the system will notify the user of a cellphone number.

2.3 USED COMPONENTS

The following components are employed in our proposed system for the SMART CHAIR in order to achieve the needed flow. The suggested model can be divided into two pieces for your consideration. The first one is for the sensor of temperature and heart rate along with the GPS, which is seen in normal ambulances. Table 2.1 lists the required components.

TABLE 2.1 System Specification

Components	Specification
Temperature measure	LM35 sensor
Heart rate measure	Pulse rate sensor
Position detection	Ultrasonic sensor
Management IDE	Arduino ESP 32
Adapter	Adapter 24 V/1.5 A
Cloud service	ThingSpeak
Bluetooth	HC-06 module
GPS	GPS NEO 6M
GSM system	GSM 800A
Software used	EAGLE (PCB design) Arduino IDE

In our proposed model of the first stage, we have used ESP 32. This module is bigger as compared with the ESP8266-01 and it is easy to use as many pins are broken out because the IO pins are facing each other, and this is very useful. For the second stage, we also employed an Arduino UNO R3 controller to manage the overall components of our model, which are all set in their positions. This microcontroller is capable of dealing with both digital and analog sensors at the same time. Another task that it is capable of performing is the transmission and reception of internet data. Next, we came up with the sensors that will be used in our model, which will include a temperature sensor, a pulse sensor MQ135 sensor, and an ultrasonic sensor (which will be discussed later). All these sensors work for a specific task, temperature sensing by LM35 sensor, Heart rate analysis through pulse sensor.

2.4 PROPOSED MODEL

As the SMART CHAIR system works in such a way that as a person is sitting on a chair, two ultrasonic sensors find out the posture of the person, and using two switches placed on the arm of the chair are used to find out the body temperature and heartbeat of a person. This model is provided with a power supply, shown in Figure 2.8.

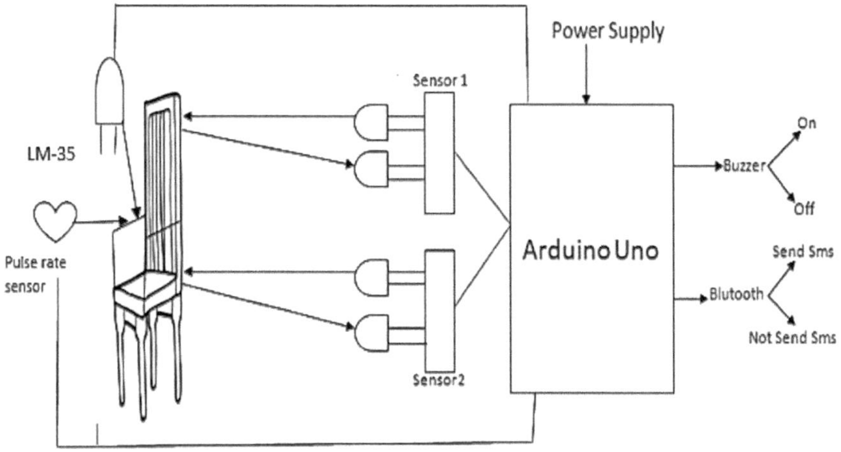

FIGURE 2.8 Proposed model.

The architecture of the system is depicted in Figure 2.8. Here we have used two Ultrasonic sensors, Ultrasonic 1 is placed on the top of the backrest, and another is placed at the bottom of the backrest above the seat. Ultrasonic sensors are used to detect the object placed before it. So, in our system, if both sensors are not detecting, then the chair is empty. If anyone sensor is detecting, then the person is either leaning or slouching on the chair. If both the Ultrasonic sensor is detecting, it means that the person is sitting with a straight posture. The Arduino UNO is linked to both of the sensor modules. Wrong posture is indicated through a buzzer, and if it is the right posture, then after every 2-hour user will get a notification on their phone to get rest. For sending a notification as an SMS, we have used a Bluetooth module and a GSM module. Pulse Sensor and Temperature Sensor are placed on the handrest of the chair. When a person places his hand on the handrest of the chair, and if the temperature or pulse rate goes high, then a message is sent to the doctor.

2.4.1 SYSTEM WORKING

- Whether people are seated straight or not is determined by their posture.
- If not, then the buzzer/speaker while indicating a wrong posture.
- If yes, then after every 2 hours, an SMS is sent to take a rest.

- LM-35 is used to check the temperature, and if it goes high, send a message to the doctor.
- The pulse sensor is used to check the pulse rate and if it goes high send a message to the doctor.

2.5 EXPERIMENT AND RESULT ANALYSIS

1. **Posture Detection:** Ultrasonic sensor 1 is placed on top of the backrest and Ultrasonic sensor 2 is placed on the center of the top of the seat. If both sensors are detecting objects, that means a person is sitting with the correct posture. If not, then either person is slouching or leaning toward the computer. The buzzer will indicate the wrong posture.
2. **Sending SMS:** If both the sensors are detecting, then after 2 hours, send the SMS using the Wi-Fi module to the user's phone to take a rest.

Table 2.2 shows results that after sitting on the table, it can detect body posture, temperature, and heartbeat successfully. This smart chair can be used in different areas like the software industry, physio-clinic, hospitals, etc. (Figure 2.9).

TABLE 2.2 Result Analysis

SL. No.	Scenarios	Expected Result	Actual Result	Status
1.	Check whether ultrasonic sensor 1 is working properly or not	Successful	Successful	Success
2.	Check whether ultrasonic sensor 2 is working properly or not	Successful	Successful	Success
3.	Check whether Bluetooth module HC-06 is working properly or not	Successful	Successful	Success
4.	If anyone of the ultrasonic sensor is detecting an object, the buzzer is working or not	Successful	Successful	Success
5.	Check whether Bluetooth modules can send SMS to the user after 2 hours or not	Successful	Successful	Success
6.	Check whether the pulse sensor is giving reading or not	Successful	Successful	Success
7.	Check whether LM-35 is giving temperature or not	Successful	Successful	Success

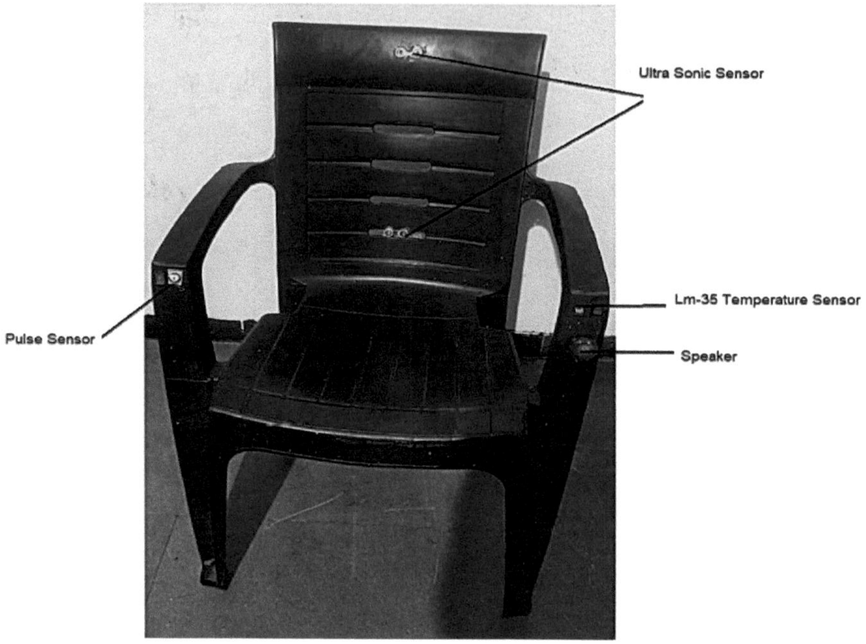

FIGURE 2.9 Smart chair.

2.6 CONCLUSION

To overcome the problem related to the human body, such as back pain, body pain, etc., which are developed due to constantly leaning towards computers also, it is noticed that sitting for so long could also change the shape of the spinal column into a curve; therefore we proposed a system Smart Chair. The newly created Smart chair technology is capable of monitoring the user's sitting behavior and assisting him or her in correcting their posture. The buzzer/speaker is used to indicate a wrong posture. Sitting with the right posture could reduce back problems. Also, the system can notify the user with an SMS to take a rest after every 2 hours of sitting because sitting straight constantly could also result in back pain.

KEYWORDS

- **Arduino board**
- **Bluetooth module**
- **component**
- **heart rate sensor**
- **integrated development environment**
- **radio frequency identification**
- **temperature sensor**
- **ultrasonic sensor**

REFERENCES

1. Kolhatkar, S., & Joshi, A. K., (2017). Automatic temperature control technique for a clinical water bath. *Proc. - 2 Int. Conf. Comput. Commun. Control Autom. ICCUBEA 2016*, 1–4. doi: 10.1109/ICCUBEA.2016.7860141.
2. Zhu, C., et al., (2015). Dual-frequency ultrasonic washing machine for fruits and vegetables. In: *2015 IEEE Int. Conf. Consum. Electron. - Taiwan, ICCE-TW 2015* (pp. 152–153). doi: 10.1109/ICCE-TW.2015.7216828.
3. Vanjale, R. S. B., & Sayalee, D., (2016). *IoT Based Traffic Signal Control for Reducing Time Delay of an Emergency Vehicle Using GPS* (p. 5). India. IEEE ISBN:978-1- 5090-0396-9/16/$31.00 ©2016 IEEE.
4. Dheeraj, D., Jitin, T., & Sarfaraz, M., (2015). *A Smart Traffic Solution for High Priority Vehicles*. ISBN: 978-1-4673-6809-4/15/$31.00 ©2015 IEEE.
5. Madisa, M. K., & Joseph, M. K., (2018). Android and cloud based traffic control system. In *2018 International Conference on Advances in Big Data, Computing and Data Communication Systems (icABCD)* (pp. 1–4). IEEE.
6. Faldu, P., Doshi, N. & Patel, R., (2019). Real time adaptive traffic control system: a hybrid approach. In *2019 IEEE 4th international conference on computer and communication systems (ICCCS)* (pp. 697–701). IEEE.
7. Vidya, B., & Ragha, L. K., (2018). *Intelligent Traffic Control System, 8*(2). ISSN 2250-3153.
8. Sarkar, A., (2020). Design of automatic hand sanitizer with temperature sensing. *International Journal of Innovative Science and Research Technology, 5*(5). ISSN No:-2456-2165.
9. Ankit, J., Lalit, K., Mayur, S., Shyam, S. J., & Sarita, C., (2015). An advance intelligent ambulance with online patient monitoring system. *IPASJ International Journal of Electronics & Communication (IIJEC), 3*(4). ISSN 2321-5984.

10. Park, M., Song, Y., Lee, J., & Paek, J., (2016). October. Design and implementation of a smart chair system for IoT. In *2016 International Conference on Information and Communication Technology Convergence (ICTC)* (pp. 1200–1203). IEEE.
11. Jayapradha, S., & Vincent, P. D. R., (2017). An IoT based human healthcare system using Arduino UNO board. In: *2017 International Conference on Intelligent Computing, Instrumentation, and Control Technologies (ICICICT)* (pp. 880–885). IEEE.
12. Snehlata, Y., & Poornam, S., (2016). Smart wheelchairs – A literature review. *International Journal of Innovative and Emerging Research in Engineering, 3*(2).
13. Hata, Y., Yamaguchi, H., Kobashi, S., Taniguchi, K., & Nakajima, H., (2008). A human health monitoring system of systems in bed. In: *Proc., IEEE Int. Conf.,* (pp. 1–6).
14. Lodhi, D., Vats, P., Varun, A., Solanki, P., Gupta, R., Pandey, M. K., & Butola, R., (2016). Smart electronic wheelchair-using Arduino and Bluetooth module. *International Journal of Computer Science and Mobile Computing.*

CHAPTER 3

Smart Farming Using Blockchain Technology: An Indian Perspective

ANJANA PANDEY,[1] ABHISHEK DUBEY,[2] BHAVESH SHAH,[3] and MUSTAFA KASARAWALA[3]

[1]*Associate Professor, University Institute of Technology RGPV, Bhopal, Madhya Pradesh, India*

[2]*Salalah College of Technology, Oman*

[3]*University Institute of Technology RGPV, Bhopal, Madhya Pradesh, India*

ABSTRACT

India is an agriculture-based country, and a large population is associated with agricultural activities for livelihood. The agriculture supply chain involves farmers, traders, mediators, local government officers, transporters, financiers, etc. The setup of physical 'Mandis' and similar infrastructures used to connect farmers and traders to sell and buy crops play a vital role in the agricultural ecosystem. However, this process is often very inefficient as each participant requires maintaining their records and ledgers at their local level, which consists of many ambiguities and flaws. Today in the era of the digital revolution, we need to adopt the use of Information Technology to make this process more efficient, trustworthy, and easy. From India's perspective, a revival of this process with the help of blockchain technology can provide a tamperproof, fault-tolerant, and universal ledger to record valuable transactions of buying and selling crops, which can have a huge impact on the lives of our farmers. For using

Intelligent Sensor Node-Based Systems: Applications in Engineering and Science, Anamika Ahirwar, Piyush Kumar Shukla, Prashant Kumar Shukla, and Ruby Bhatt (Eds.)

any digital solution, the biggest challenge is always to fix the trust gap between the participants who interact virtually without seeing or physically being with each other. There is always a sense of ambiguity and a fear of fraud during online transactions. Blockchain technology addresses this trust gap and allows the user to trust the ledger since it is immutable blindly. No one can tamper with the records once they are stored on the ledger and visible to everybody.

3.1 INTRODUCTION

We are living in the age of globalization, where all parts of the world are easily approachable, and the whole world has become a market. Therefore, we can consider the whole world as a client, and we can improve our products based on the demands. Demand and supply have always been a great concern for all countries, and aligning the production of crops with their demands can change the whole scenario. The agricultural market, which is mostly considered a nonprofit or minimal profit-taking area, can become the most profit-generating field. Our farmer's life can be drastically changed, which is mostly all countries' central theme of their policy (Indian Agriculture Policy) (China's Agriculture Policy) (WTO) [1]. As it is stated that we are living in the age of globalization, and most of our lives the aspects are connected with the digital world but still, our farming is away from it, though there are a lot of IoT devices being invented and research going on in all aspects of the agriculture industries, to improve the production, optimize the use of land, etc., still the global traders and farmers are not linked with each other, they are not aligned. What are the trader's demands, what are the farmer's capacity to produce the crops, what is the land's quality to produce without much effort, and does the particular environment of the region supports the crops to produce in their fullest? These are all factors when they come together and align with each other, and then they may create a revolution in agriculture industries and also pave a path of prosperity in the whole world where food will be available for all and ANNDATA (Farmers) will always be happy so that all people may also enjoy the prosperousness? This dream can be only possible when all actors come together and make an effort collectively. To support this, we need a technology that can provide a platform where traders and farmers will be together linked with all government and non-government agencies in a trustful environment. The blockchain is a technology that can fulfill the entire mentioned criterion.

In this chapter, we have used blockchain technology to link the traders and the farmers, and other actors. Based on the demands raised by the traders, the farmers will be able to produce the crops. This proposed method will provide farmers and traders security while production and trading. Optimizing agriculture production has always been a central theme of all the agriculture policies of all agriculture-based countries. Due to many reasons, it was not materialized up to now. Through the proposed method, we can provide approximate demand of various corps from across the globe to the farmers and predictions about the production to the traders. It will also strengthen the relationship between both of them. This step will further increase the implementation of the new technology in the agriculture industries and all actors such as Traders, Farmers, Government agencies such Meteorological Department, Soil health department, Agriculture Department, Land reform Department, Banks, Auditing departments, international institutions, Transport Companies, Chemical fertilizer industries, insurance agencies, and all related industries to work together to achieve a goal. It will align all the actors to work towards an inclusive force for a collective objective. This proposed model will improve the supply chain model and eliminate/minimize the role of mediators.

In India, since its independence government has been trying to create a cooperative culture between farmers but could not succeed due to many reasons. Our model gives an opportunity to create cooperative groups intentionally or unintentionally means based on the demand and minimizing the risk the system can form a cluster of farmers to produce certain crops. It will be a naturally cooperative society without any force and also gives an opportunity to minimize the risk of natural calamities through the creation of clusters of producers from different geographical parts. Farmers may officially form cooperative groups and participate as a unit in this model. Our model will also encourage corporates to invest directly in farmers instead of owning land and investing their major portion of liquidity in corporate farming, which may have many risks.

From the governance point of view, this proposed model can play a vital role. Currently, due to lots of flaws in our present system, farmers have to suffer, for example, the claim of insurance for their production. As per the system, farmers have to produce so many documental proofs, and along with it, so many old rules restrict them from being benefitted from their rights. If that particular area's a certain portion of farmers will not claim it, then also farmers will not be able to get appropriate benefits and receive the actual loss through insurance. Though the government's intention is positive, we do not have an efficient system that could help the government to assess the loss of a particular farmer. This aspect and many other aspects can be added to the current model and easily implemented policies that can help the farmers and bring their confidence back in agriculture.

From the Indian perspective, the farmers are the producers of various crops, which they sell either through government-supported agencies or mediators.

This process of selling and buying crops is very inefficient because each participant in the process, such as mediators and officers, maintains their ledgers to keep records of all the transactions, which leads to more corruption, inconsistency, and errors. This problem can be solved with the help of blockchain technology. Today as most of the data is being stored digitally; blockchain provides a digital ledger platform that is immutable, fault-tolerant, shared, and transparent. This provides a trustworthy and universal ledger that can be used by each participant in the agricultural supply-chain network to record transactions.

With the use of blockchain technology, the whole food supply chain can be made more transparent, and more information can be made accessible to all the actors involved in the agriculture supply chain. The trader can easily have access to all the information about the crops, farmers involved in the production of the crop, pesticides used, land used, and various other kinds of information. The consumer of the food can get to know all the details related to the food item he is purchasing by systems such as scanning the barcode on the food item.

In this chapter, we have provided a proposed model for farmers and traders (Flow charts are attached). An empirical study of the current model and the prospects of the proposed model along with advantages and challenges, how our proposed model can make a change and become beneficial for the government, farmers, traders, and other related sectors or industries related to agriculture have also been provided.

3.2 EXISTING SYSTEM AND PROBLEM

The food supply chain consists of numerous actors who play their roles in this complex system of food supply right from the beginning the farmers who produce the crops, traders who buy their yields, distributors, shipping, and transporting agencies, local governing bodies to regulate the process, cold storage companies, insurance companies, grocery stores, supermarkets and finally the end consumers. The government also plays a major role by formulating various schemes for the benefit of farmers, providing them insurance for crop damage, providing food security, food safety, and food integrity, observing transactions taking place in the market, and resolving conflicts in the process [2].

In the currently adopted system, all the actors involved in the agriculture supply chain maintain their ledgers, and the local authorities responsible for regulating the process maintain their copies of records and ledgers. Thus, there is a lot of ambiguity in the whole process, and a great deal of corruption as the individuals maintaining their ledgers can easily temper with the data and records for their advantage and personal gain. This makes the whole system corrupt. The agriculture supply process works on such a large scale, and all the actors are distributed and perform their activities in their regions. It gets impossible for the government to bring transparency, reachability, and regulate the whole process in an efficient manner. Ultimately the farmers, traders, and government are left with no choice but to trust those ledgers maintained at various stages of the agriculture supply chain. Blockchain technology can completely solve this trust problem and will allow the government to win the trust of farmers and traders. The blockchain [3] will provide a secure and safe platform for maintaining a common ledger storing all the information and records throughout the process accessible to all.

3.3 PROPOSED MODEL

3.3.1 FOR GOVERNMENT

The government plays a major role as a central authority governing and regulating all aspects right from the beginning of crop production to its being sold at grocery stores, vegetable markets, and supermarkets to the end consumers. In this model using blockchain technology, the government's role would be to provide a safe and trustable channel for trading crops, to closely observe the transactions taking place between various actors involved in the process, and to resolve any complaints and conflicts if they arise at any stage of the food supply chain.

The government will benefit from this model by utilizing its transparency, accessibility, fault tolerance, and immutability. As the blockchain provides a common platform for all the actors involved in the food supply chain, such as farmers, traders, shipping companies, groceries, distributors, cold storage companies, and insurance companies, to maintain their records and data on a commonly distributed ledger. It thus provides the government an ideal platform to implement and execute all its schemes and yojna in a very transparent, clean, and trustable manner removing all the intermediaries, middlemen, local officers, local authorities, and local bodies from the process. All the records and details of every stage and every actor involved in the whole process will now be stored on a single digital ledger accessible to all. The government will now directly communicate and implement its schemes through this platform, and all the information and record will be transparent and visible to everybody. The blockchain provides ultimate security and nobody can mutate or temper any data stored on the blockchain. All the computation and recording of new transactions take place on the blockchain through the consensus of a large network of nodes executing the smart contract on a virtual machine running on a distributed network.

The model will be implemented such that the government will host a website that will act as an interface between the public blockchain and its users, who are the farmers, traders, shipping companies, distributors, cold storage companies, groceries, end consumers, and the government itself. All the participating actors will access the blockchain by registering and making their accounts on this government portal. This portal will provide the interface for performing various actions, and on the backend, all the records will be stored on a public blockchain. The backend logic of the website will be built as a smart contract which will be deployed on the public blockchain. The government will now just need to maintain the front end of the website, and the smart contract, once deployed on the blockchain, will now be executed by the blockchain network. All the computation and storing of the data takes place in the blockchain network itself, so the government will not need to maintain its data centers or purchase services for this purpose anymore. The government will invest in the blockchain and will pay for all the transactions that will take place on it as the blockchain itself does not work free of cost and the miners get their proportion of money. The government can also launch a smartphone app and can also use internet of things (IoT) technology to bring ease in recording new transactions (Figure 3.1).

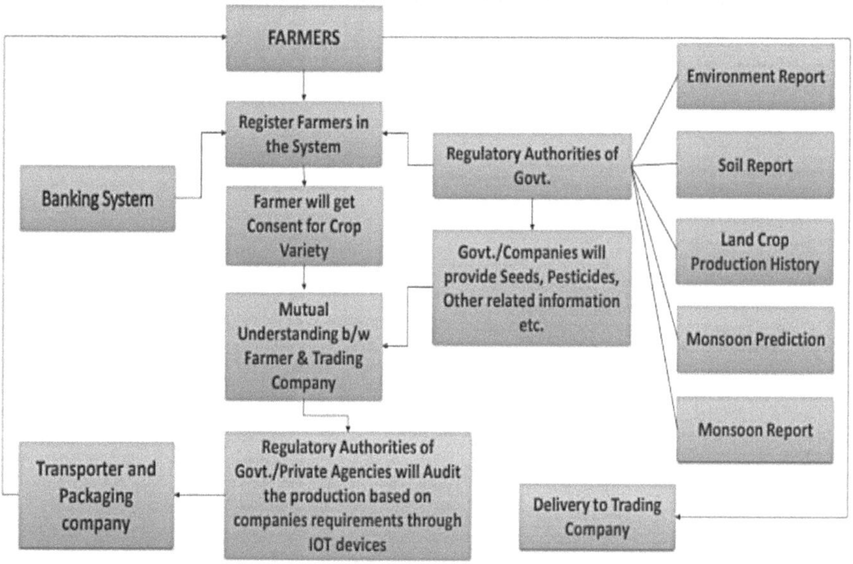

FIGURE 3.1 Government model.

3.3.2 FOR FARMERS

Farmers will register through government institutions such as Gram Panchayat/Cooperative Bank/Samiti/Mandi. The farmer will register through the local regulatory authority. Each registered farmer will be issued a ledger in which they have to enter details of their crop production details. The farmers will need to fill in their details, such as the amount of their yield. The types of crops they have produced, the location of their farming land, agricultural methods, and practices used fertilizers, and pesticides used, the quality of their crop, the yield produced in previous seasons, and the total expenditure, income, and losses in their corp. If any damage is caused to their crop, they need to fill in the details of their losses and the degree of damage caused, the cause of and category of damage, and the sum of losses. The government can also implement IoT technology to monitor for such parameters and completely automate the process causing the farmers to fill in only basic details, and the rest will be will automatically recorded by various devices such as drones, water level indicators, soil quality meters, etc. Since they will be connected through the internet, new transactions will be automatically communicated to the blockchain without any human effort (Figure 3.2).

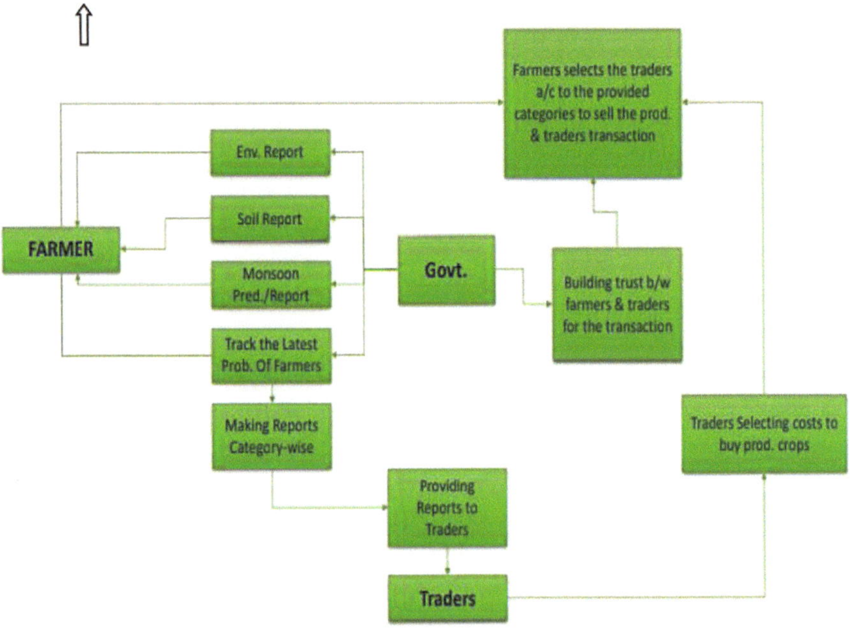

FIGURE 3.2 Farmers model.

3.3.3 FOR TRADERS

Trading companies will be able to register through government regulatory authorities. Each registered trading company will be issued a ledger in which they have to enter details of their requirements and needs, the details such as when they need to be delivered, the amounts of crops and the types of crops, and what are their other demands.

The traders will get the ability to perform various searches and market surveys with the help of the blockchain sitting in their offices. The portal will allow them to perform various queries and access information about the details of various crops, total yields, availabilities in their nearby regions, the details of the farmers, the complete details of the crops, etc.

This platform will not only provide a secure medium of transactions and accessing information but will allow them to directly communicate with the farmers removing all the middlemen and intermediaries (Figure 3.3).

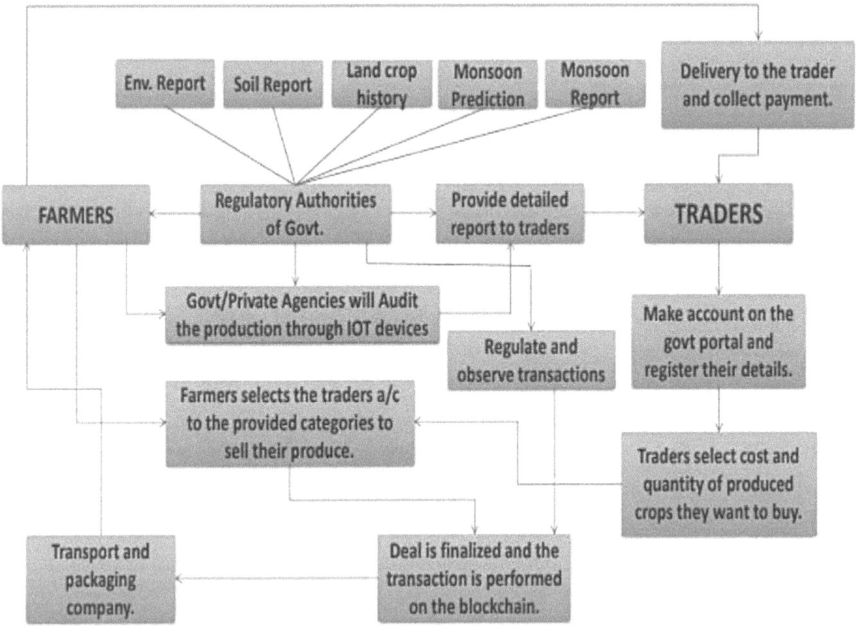

FIGURE 3.3 Traders model.

3.4 HOW BLOCKCHAIN TECHNOLOGY IS BEING USED TODAY?

Blockchain is an emerging technology that works on a distributed ledger system providing fault tolerance, transparency, and ultimate security for handling and storing various transactions and records. Blockchain is being used for providing a trusted and secure platform to perform various activities such as maintaining administrative records and enabling electronic voting to be secure and tamper-proof. Blockchain is being used as a ledger for cryptocurrencies, maintaining health records, and for non-fungible tokens to sell and buy digital assets on the internet, online auctions, online donations, real estate, etc. Blockchain provides consensus and a sense of trust when multiple stakeholders or entities, or businesses are collaborating, such as in various industrial supply chains. In the agriculture sector, blockchain technology has a great scope and potential to solve various problems occurring in the whole agriculture supply chain due to the chaotic nature of the process and the unorganized management of information.

3.5 HOW EXISTING MODELS ADDRESS THE PROBLEMS IN AGRICULTURE SECTOR?

On the detailed study of many existing models currently working in various regions across the globe, we can observe that most of them are more concerned with traceability and transparency in the food supply chain. That is by recording and maintaining all the details of an individual product and tracking its journey right from the farm where it is produced to the grocery store where it ends up in the customer's hand as a final consumable product. These models focus on capturing all the essential information associated with a particular product, right from the pesticides and fertilizers used in the farm to the warehouse and cold storage centers where it was stored to the manufacturing plants where it was processed, and finally to the store where it was finally kept before the customer purchased it.

For food producers, adopting the use of blockchain technology helps them to win the trust of their end customers. For example, a dairy milk selling company decides to put a QR code on the package of their product which can be easily scanned by any supermarket customer using a smartphone app. On scanning the code, the customer gets access to a website that displays all the transactions recorded on a blockchain. From here, the customer can easily track the details of the particular item he is purchasing, such as when it was produced, from which cow, how fresh it is, where it was stored, which quality tests were performed on it, etc. The customer can visit the complete timeline of the product right from its origin on the farm to its ending up on a supermarket shelf. Since all this information is recorded on an immutable, transparent, and universal ledger provided by the blockchain, the customer can trust the information to be 100% genuine. This will help the producer to establish a trustful relationship with its customer and will help in the branding of their product, that they sell only genuine and fresh products. This can help a business stand out in the market and achieve growth.

The existing models also address the problems and conflicts that arise during agricultural land registrations. Various initiatives are being taken by governments to make the land registry more reliable [4]. Blockchain technology can be deployed here to record the land property ownership and transfer details. This will be beneficial for farmers as their farming lands can be linked to their sovereign ID/digital ID. It will help in safe-keeping and securing their lands and settling land disputes. The Swedish government's [5] land-ownership authority, Lantmäteriet, has prepared a model to record land registry and property transactions on the blockchain, and a lot more work is under process which may revolutionize the human beings, society, and the whole world as well.

3.6 EXAMPLES OF SOME EXISTING BLOCKCHAIN-BASED MODELS

3.6.1 BARILLA, ITALIAN PASTA, AND PESTO SAUCE MANUFACTURER, ITALY [6]

Barilla is an Italian multinational food company and is one of the largest producers of pasta and pesto sauce in the world. Along with IBM, it has prepared solutions to improve transparency and traceability in its pesto production cycle. All the details of its production, right from cultivation to processing to storing to transporting information, are tracked and recorded on a blockchain-based IBM cloud infrastructure and made available to the customers who can easily verify all these details of production by scanning the pesto's QR code.

3.6.2 AGRIDIGITAL, AUSTRALIA [7]

AgriDigital was founded in 2015, and since then, it has been working with blockchain to provide reliability and traceability in the food supply chains, primarily in the grain supply chains. This startup provides a cloud-based commodity management solution platform to its users, which has been built to be blockchain-enabled. This platform primarily acts as an interface between the user and the blockchain layer. It is a multi-participant commodity management platform and joins farmers, buyers, financiers, and end consumers with each other to interact through a single platform.

3.7 HOW OUR MODEL IS DIFFERENT THAN THE EXISTING MODELS?

The existing model aims to build a trusting relationship between producers and end consumers of agricultural products by providing traceability and transparency in the food supply chain. This ensures the food quality, builds its reputation in the market, and strengthens the trust of consumers in the product. Whereas the model suggested by us, the aim is to build a trusting relationship between the producers of various crops and the traders who purchase from them. The model does not focus to benefit the end consumer of the products instead it focuses to benefit the producer, i.e., the farmer. Though this model has a positive impact on all actors. The model aims to address the communication gap between the farmers and traders, by providing a secure platform where farmers and traders can interact, settle prices, and transact with each other and the government can play its role of regulating and mediating those transactions of crops effectively.

3.8 ANALYSIS

3.8.1 HOW THIS MODEL CAN BE BENEFICIAL FOR FARMERS?

This model would help farmers to get rid of the traditional paper ledger system to maintain crop production and export details. The security of all the information is guaranteed by Blockchain. The Credit Score system in the model would help in producing better throughput of crops. The insurance facility of this model would help farmers, not to stand in a queue or go to a nearby city in order to prove their eligibility, because all the necessary details will be available in a distributed ledger.

This model will make a productive environment for farmers and the industry as well, and it gives an opportunity to link the food industry and farmers directly without any middleman. Government can govern the whole process and eliminate the hurdles of the farmers and provide direct help and benefits to them without any biasedness. The three amendments which have been done by the government will pave a path for this kind of proposed system and endorse our concept behind proposing this model.

The local bodies and middlemen responsible for providing the benefit of various schemes formulated by the government for the benefit of farmers are mostly involved in corruption and play with the records to keep the money allotted for subsidies and help farmers themselves; as a result, the farmers are unable to benefit from these schemes. The blockchain will bring transparency to the system, and it will get impossible to tamper or mutate the records.

3.8.2 HOW THIS MODEL CAN BE BENEFICIAL FOR TRADERS?

This model could impact trading with farmers in several ways. If the trades of crops were recorded on a blockchain ledger, settlement time could drop from days to minutes, with improving efficiencies and flawless processes. This distributed ledger would help in potentially reducing the cost. The industry can raise their needs and communicate to the producers about the market needs, which will create a stable market driven by actual needs and requirements instead of the current market, which is driven by speculations and uncertainty of demand and supply.

3.8.3 HOW THIS MODEL CAN BE BENEFICIAL FOR THE GOVERNMENT?

3.8.3.1 HOW THIS MODEL WILL HELP FCI TO COORDINATE AND PLAY ITS ACTUAL ROLE?

The Food Corporation of India is the statuary body under the Ministry of Consumer Affairs, Food and Public Distribution, Government of India. The decentralized ledger maintained throughout the process would help FCI to keep an eye on all the activities, including trades, production, supply, etc. Breaking the integrity of the system will be the most difficult task. Even if any fraud in the process occurs, it would be easy for the ministry to backtrack and find out where the malfunctioning in the process took place.

3.8.3.2 HOW MEDIATOR'S ROLE WILL BE LIMITED?

The model would help all three major bodies of the system, i.e., Farmers, Traders, and Government to stay in direct contact. It aims to minimize the role of mediators by maintaining an open and secure ledger. The current Scenario involves mediators between Traders and Farmers, Farmers and

Government. The introduction of this model would help in saving the cost that is given to mediators.

The Government's vision to double farmers' income can become a reality even based on these models, and the farmers can get their products' actual valuation.

3.8.3.3 HOW THIS MODEL WILL BE HELPFUL TO STOP THE BLACK MARKETING OF FOODSTUFFS?

The blockchain-based model focuses on the secure generating of records and bringing transparency throughout the supply chain, to reduce food fraud and enhance food safety. Integrity is a major concern here, i.e., if 100 kg of the crop is ordered, 100 should reach to the destination. If any malpractice occurs in transportation, that could be easily tracked. This would also help Government in strengthening the country's economy.

3.9 CHALLENGES (LIMITATIONS)

1. The various actors involved in the food supply process, such as farmers, traders, and distributors, need to be motivated to provide genuine and precise information to the blockchain ledger.
2. The farmers, traders, and local individuals who act in the food supply chain are mostly not educated to operate on computers and use the internet. They will find the digitalization of the record-keeping and using digital ledgers a more anxiety-prone task and will not be able to take complete advantage of the model.
3. There would be a risk of fraud as those farmers and traders who don't have intricate knowledge of using the internet might accidentally land on a fraud clone website, causing them to become victims of fraud transactions hosted by that website.
4. Poor internet connectivity and lack of infrastructure in remote rural areas would make it difficult for farmers to access the platform.

3.10 RESULTS AND DISCUSSION

When the user opens the platform in his browser window, he is basically interacting with the front-end web application, which is built using JavaScript, HTML, CSS, and bootstrap framework and is hosted on a web hosting service or a cloud. This website plays the role of interface between the user and the blockchain. The blockchain at the back end can be a permission-less public blockchain platform that is programmable; that is, the developers can write logic into the blockchain in the form of smart contracts (Figure 3.4).

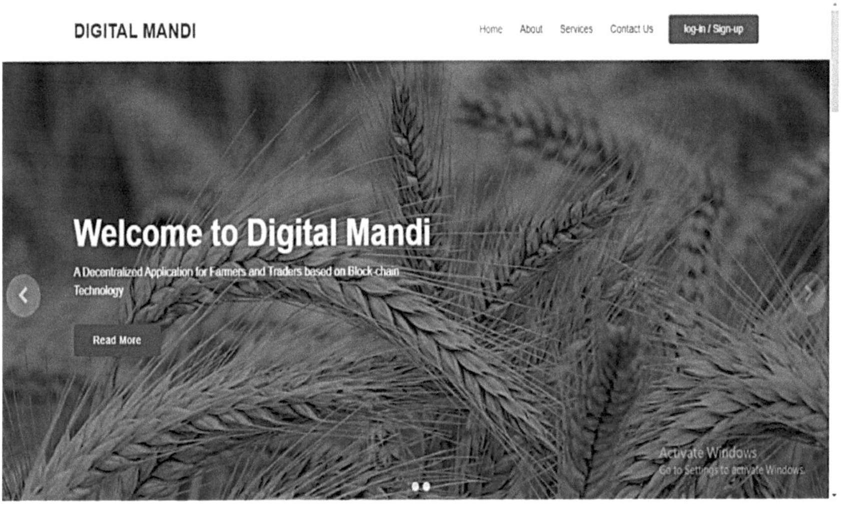

FIGURE 3.4 Home page of the decentralized web application.

In the front-end module, a special service named "BlockchainService" has been defined, which would link the complete front-end with the backend blockchain. This service would import the relevant library for the specific blockchain platform. This library acts as an API for interacting with the blockchain and contains an object called "Contract," which represents the actual smart contract stored on the blockchain and allows us to call the functions of the smart contract and interact with it. In the "Blockchain-Service," a function "ExecuteTransaction()" has been defined, which in its definition calls the functions of the smart contract with the corresponding arguments. Each time the "ExecuteTransaction()" function is called, a new transaction is made on the blockchain. A new transaction is generated each time the user performs certain activities on the website, such as registering a new user, registering the details of the product, making a query about crops, making a purchase request for a crop, accepting orders, making payments, acknowledging delivery, requesting statistics and graphs about crop production, defining new constraints via admin mode, etc. For each new transaction, the "ExecuteTransaction()" function is called, and the details entered by the user and his blockchain public address key are passed as parameters to the function. For example, if a farmer registers himself on the website and enters his details and ID number and the details of the produce he wishes to sell. All this information and the farmer's public key address will be passed as parameters to the "ExecuteTransaction()" function, which in turn would call the "RegisterFarmer()" function of the smart contract.

The smart contract represents the core logic of the project. Instead of having a centralized server in the back end, all the transactions are stored on a decentralized ledger and processed on the virtual machine running on a network of miner nodes. The smart contract serves as the backend logic that defines how the new transactions will be validated, which data structure will be used to store data, how data will be mapped with users and how data will be fetched from the blockchain. The smart contract itself is stored on the blockchain and executes on the virtual machine of the blockchain. In the smart contracts, we have defined classes for farmers and traders and functions which define how the objects of these classes will interact with each other and how data will be stored and processed. The functions defined on the smart contract contain those Boolean expressions that determine whether a transaction is legal or not. These pre-defined terms and conditions will prevent any fake or illegal transaction from happening on the platform. Some of the primary conditions that have been defined in the smart contract are:

- The farmer cannot sell the same product multiple times to different traders. That is, once a farmer has sold his crop, and the transaction has been verified and recorded on the blockchain, the farmer cannot commit fraud by selling it again to some other trader, thus causing inconsistency in the database.
- The trader cannot underpay a farmer, and similarly, the farmer cannot demand an excessive amount for his crop. That is, the buying and selling of any particular crop can only happen in the legal range of price decided by the government.
- The farmer or trader cannot provide any fake information on the platform. For example, a farmer is trying to commit fraud by filling in fake information about his crop on the platform and not delivering the same to the trader. In this case, since all the

transactions and activities are being recorded on the blockchain, they cannot be tampered with, and all the activities can easily be traced. Thus, helping the authorities to track all the transactions of the farmer and take appropriate action against him. Similarly, a trader trying to cheat by doing fake transactions can be easily tracked. The process can be made more trustworthy and accurate by using IoT devices and sensors to capture the details to be stored on the blockchain.

3.11 CONCLUSION

Blockchain is still in-progress technology, but it has the power to make the system secure at the highest level possible to date. Its application in the farming sector can bring a revolutionary change in digital India. Many technology experts believe that Blockchain technology can create the same revolutionary impact as the Internet did twenty years ago. Therefore, keeping in mind the integrity of information, fraud detection, and profits of farmers and traders along with the government of India, this model can give us many miraculous outcomes.

KEYWORDS

- **agriculture**
- **blockchain**
- **farmers**
- **government**
- **traders**

REFERENCES

1. Hang, X., & Tobias, D., (2020). *Blockchain Technology for Agriculture: Applications and Rationale*. https://www.frontiersin.org/articles/10.3389/fbloc.2020.00007 (accessed on 21 December 2022).

2. Geard, S., (2019). *Blockchain for Agriculture Opportunities and Challenges*. http://www.fao.org/3/CA2906EN/ca2906en.pdf (accessed on 21 December 2022).
3. Shyamala, D., & Suguna, R., (2019). Design of IoT blockchain based smart agriculture for enlightening safety and security. *Emerging Technologies in Computer Engineering: Microservices in Big Data Analytics* (pp. 7–19).
4. https://www.undp.org/content/undp/en/home/blog/2018/Using-blockchain-to-make-land-registry-more-reliable-in-India.html (accessed on 21 December 2022).
5. https://cointelegraph.com/news/sweden-officially-started-using-blockchain-to-register-land-and-properties (accessed on 21 December 2022).
6. https://bitnewstoday.com/market/blockchain/icons-of-italian-business-opt-for-blockchain/ (accessed on 21 December 2022).
7. https://www.futurefarming.com/Tools-data/Articles/2019/5/Agridigital-to-create-digital-trust-in-supply-chains-424208E/ (accessed on 21 December 2022).

CHAPTER 4

An Intelligent Parenting System: Artificial Mother Monitoring System for Sleeping Infant

HARSHITA JAIN

Department of Computer Science and Engineering, SIRT, Bhopal, Madhya Pradesh, India

ABSTRACT

With the advent of the internet of things (IoT), artificial intelligence (AI), and machine learning, the life of parents has become easy. Meanwhile, the responsibility for taking care of their children has also become easier now. As we know, in this era of Technology and Crime, the rate of child abuse and kidnapping is increasing day by day. So, the safety of their children becomes one of the major concerns for the parents. IoT technology can help out parents to take prior security solutions, including location tracking capability, and prevent such crimes from happening. According to research conducted by the University's Genpact Women's Leadership Center, 50% of professional women in the country leave their jobs at age 30 to take care of their children. Even among the women who managed to do so, 48% quit within four months of re-entering the job market. Hence proposing an Intelligent Parenting System can help out mothers who are working or wants to keep working after having a baby. The use of this technology for smart parenting can be included even in sorting and solving the never-ending queries of children, for the self-assessment of children, and also can help entertain your child. On the other hand, IoT technology

Intelligent Sensor Node-Based Systems: Applications in Engineering and Science, Anamika Ahirwar, Piyush Kumar Shukla, Prashant Kumar Shukla, and Ruby Bhatt (Eds.)

can help out mothers even during the infancy of a baby child. During the initial months of a newborn baby, the mother has to take care of the child day and night without leaving the baby alone. Thus, it generally becomes a challenge for working women, and hence they take long leaves to take care of their babies, and even some mothers quit their job. Mothers work only when their baby sleep, but mothers cannot go too far from the baby as the baby may wake up anytime and cry. So, with the help of IoT technology, the Artificial Mother Monitoring System will monitor the baby through sensors and will let the mother know if the child has to wake up or if something is not right through SMS. As soon as the mother receives the SMS, either she can assist someone else to go near the baby if she is out of home, or she herself can reach the baby if she is just a room far away from the baby. This monitoring system works on two sensors, i.e., a PIR sensor to monitor the physical movements of a baby and the other is noise sensor that detects whether the baby is crying or not. This system works on a hybrid combination of noise and PIR sensors. The thresholds will be set in a combination of both PIR and noise sensors. Sometimes baby keeps moving their head or sometimes their hand and legs too, which does not mean they are awake, so the combination of noise and movement intensity will decide whether the baby is awake or not, and once the decision is taken by the algorithm, the message will be sent to the mother. Thus, we can call this system an intelligent artificial mother monitoring system.

4.1 INTRODUCTION

The well-documented fact is that India's workforce is biased toward men and not conducive to women. Although women have very few imports, they have many exports: pregnancy, childbirth, childcare, elderly care, lack of family support, and unsupported work environment. These factors created a leaky pipeline and prevented women from occupying leadership positions. It is indicated that if the export door remains open, it will be difficult to achieve the global goal of a 50:50 ratio of the male-to-female workforce by 2030 [41]. The study stated that women expressed challenges on four levels. On a personal level, most people say they feel guilty. At the family level, they expressed a sense of "commitment." Another main reason given by women is that supervisors think they are not productive and efficient. Indian social norms require women to stay at home and take care of their children, which also plays a role. The overall condition of working women has been shown in Figure 4.1. So, the researched and proposed framework will help out the working moms and also the moms who are home alone and have to look after a baby and at the same time complete household chores.

FIGURE 4.1 Graph showing statistics of Indian working mothers [41].

4.1.1 OBJECTIVES

The objective of this research is to solve the problem of every new mom by exploring the features of the internet of things (IoT), AI, and machine learning. To achieve this aim, some research questions are needed to be answered:

- How do IoT-based baby monitoring systems work?
- What architecture can be used for artificial baby monitoring systems?
- How is the reliability of data ensured in IoT-based monitoring system systems?

4.1.2 MOTIVATIONS AND PROBLEM STATEMENT

Under relentless life conditions, everybody is occupied in their expert life, including guardians. They go out promptly in the first part of the day and return before supper time. Indeed, even the moms are working. In this way, they don't have adequate opportunities to deal with their children. Not all guardians can bear the cost of a babysitter to assist them with their kids. At that point, in the wake of working for extended periods, the moms, despite everything, need to deal with the house and deal with their children at the same time. Guardians probably won't have the opportunity to calm their infant to rest or rock their child back to rest in the night. Learns about the impact of shaking an infant have been made, and it discovered that children rest better while being shaken or swung daintily in light of the fact that the musical development mirrors the delicate shaking, they felt while in their moms' bellies. Most accessible mechanized supports are intended to shake relentlessly. Be that as it may, shaking development can make the infant sick and awkward. In this way, permitting the robotized support to shake the child to rest the night is likewise an issue. Besides, a few guardians place their children in different rooms. In this manner, guardians couldn't hear the infant crying and couldn't be there to move their child back to rest in the night. Different guardians might be busy with house tasks. In this manner, since they can't hear their infant crying, they can't take care of them right away. Once in a while, the child just needs a little interruption to come back to profound rest. A few sorts of infant supports are accessible in stores; however, they are costly, and not every person can bear the cost of them. What's more, the currently programmed supports in writing have numerous impediments as far as usefulness, cost, and correspondence innovation support. As far as we could possibly know, no past investigations have built up a keen support with IoT support without any preparation, like that in the current examination. To beat this issue, another programed IoT-based child checking framework is planned, permitting the guardians to get to a record to screen the infant's condition anyplace and whenever.

4.2 RELATED TERMINOLOGIES

4.2.1 INTERNET OF THINGS (IOT)

It's a collection of networked processing devices, mechanical and smart machines, animals, or people with interesting personalities and the capacity to transfer data over a network without requiring human-to-human or human-to-machine communication.

4.2.1.1 POWER OF IOT

The internet of things, or "IoT," is a concept that aims to extend the web's capabilities beyond computers and smartphones to a broad range of new goods, processes, and environments, rather than simply computers and smartphones. Those "connected" objects gather, transmit, or do both at the same time information.

4.2.1.2 WHY DOES IOT MATTER?

The IoT allows businesses and individuals to have a better understanding of and control over the 99% of items and situations that are not accessible through the internet. As a consequence, the IoT allows businesses and individuals to become more linked to their surroundings while also doing labor that is becoming more crucial and important in nature. When anything is linked to the internet, it implies that it can send and receive data, or that it can do both. Because of their capacity to communicate or maybe receive data, things become intelligent, and being wise is acceptable. We should, for example, reinstate the usage of mobile phones (cellular phones). Right now, you can listen to practically every music ever recorded on the planet, but this isn't because your phone has every song ever recorded on it. This is because every music on the planet is kept someplace else, yet your phone may send data (inquiring about a certain melody) and get data in return (spilling that tune on your telephone). It is not required to incorporate super stockpiling or a supercomputer in order for anything to be clever. Every task that must be completed must be linked to either a super stockpile or a supercomputer. It's a joy to be associated with you. As a consequence of the IoT, people can live and work more intelligently, and they can also have complete control over their lives. IoT is also important to companies because it enables them to market smart home automation devices. The IoT provides businesses with a real-time check on how their systems are doing, providing real-time information into anything from machine performance to supply chain and logistics operations. Businesses may use the IoT to automate processes and save human costs. It also cuts waste and improves service delivery by cutting production and shipping costs and offering greater insight into customer interactions. As a result, IoT has become one of the most important lifestyle technologies, and it will continue to gain traction as more businesses see the value of connected devices in helping them stay competitive.

4.2.1.3 CONNECTION OF INTERNET OF THINGS (IOT)

All the things that are being connected to the internet can be put into three categories (Figure 4.2):

1. **Collecting and Sending Information:** The usage of sensors is required for this. Sensors may be used for a variety of purposes, including temperature sensors, movement sensors, moisture sensors, air quality sensors, light sensors, and other sorts of sensors. These sensors, when combined with an association, enable us to get data from the ground in a natural way, enabling us to make better-informed judgments.

   By collecting data on how moist the ground is in the natural environment, ranchers may get precise information on when to water their crops on the farm. Ranchers can guarantee that crops get precisely the proper quantity of water, rather than watering excessively (which might result in costly overuse of water system infrastructure and environmental inefficiencies) or watering insufficiently (which can result in a costly loss in yields). Ranchers will make more money, and as a consequence, the world will have more food.

   Sensors enable robots to understand their environment in the same way as human senses of sight, hearing, smell, touch, and taste allowing us to experience our surroundings as people.
2. *Getting and Acting on Information:* Machines acquiring data and acting on it is a notion we're all acquainted with. Your printer gets an archive and prints it out. Your vehicle keys send a signal to your automobile, which causes the doors to open. There are an endless amount of different models to choose from. The ability to control devices from a distance, whether it's as basic as telling a machine to "switch on" or as complicated as sending a 3D model to a 3D printer, is a revelation. The actual intensity of the IoT comes when things are capable of executing both of the aforementioned activities. Those who gather and transmit data, as well as those who receive and follow up on data.

3. *Doing Both:* Returning to the cultivation model, let's take a short look at it again. The sensors can gather data on soil moisture to inform the rancher on how much to water the crops, but the rancher isn't necessary. Instead, the water system structure may turn on and off based on how much moisture is in the soil. If you want to, you can go one step further. If the water system framework gets weather data from its internet connection, it can forecast when it will rain and decide not to water the crops today since they will be irrigated by rainfall anyhow. This information on soil moisture, how much the water system is watering the crops, and how well the harvests really grow can be gathered and supplied to super-computers, which do extraordinary computations to understand the data. It's also just one kind of sensor. These algorithms might learn a lot more if additional sensors like light, air quality, and temperature are included. With a few, hundreds, or thousands of ranches gathering this data, these computations may give unimaginable insights into how to cause yields to expand at the fastest possible rate, assisting in the care of the world's growing population.

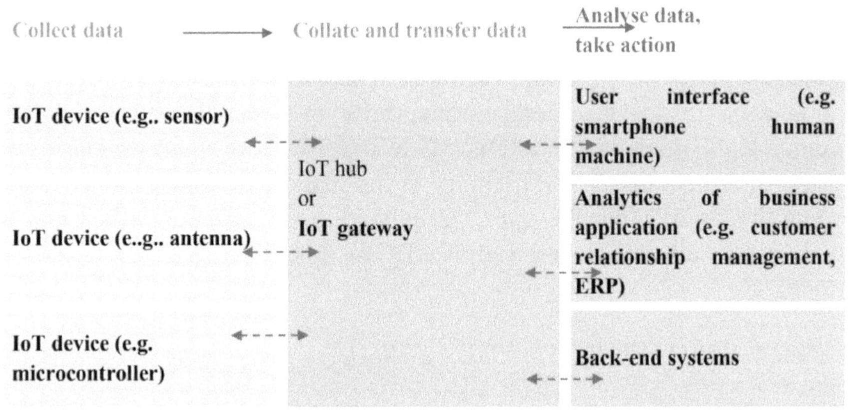

FIGURE 4.2 Flow diagram of working IoT model.

4.2.1.4 *ASPECTS OF IOT*

The following are some of the advantages of IoT:

- Enhanced communication between related electronic devices;

- Transporting data packages through an associated network, saving time and money; and
- Automating tasks, aiding with enhancing the quality of a company's administrations, and reducing the need for human intervention.

The following are a few of the IoT's drawbacks:

- As the number of connected devices grows and more data is exchanged among them, the possibility that a programmer may access classified information grows as well.
- In the long term, businesses may need to manage massive numbers of IoT devices – potentially millions – and collecting and analyzing data from each one will be challenging.
- All things considered, if there is a defect in the framework, every related gadget will be debased.
- Because there is no universally accepted IoT standard, it's difficult for devices from different manufacturers to communicate with one another.

4.2.2 SENSORS

Sensors are complex gadgets that are every now and again used to identify and react to electrical or optical signs. A sensor changes over the physical boundary (for instance: temperature, pulse, moistness, speed, and so on.) into a sign which can be estimated electrically. How about we clarify the case of temperature? The mercury in the glass thermometer extends and gets the fluid to change over the deliberate temperature, which can be perused by a watcher on the adjusted glass tube.

4.2.2.1 CRITERIA TO CHOOSE A SENSOR

There are certain highlights which must be viewed when we pick a sensor. They are as given beneath:

i. **Precision.** Closeness of measurement of readings from sensor.
ii. **Natural Condition:** Generally has limits for temperature/ dampness.
iii. **Range:** Measurement cutoff of the sensor.

iv. **Alignment:** Essential for a large portion of the estimating gadgets as the readings changes with time.
v. **Goals:** Smallest addition distinguished by the sensor.
vi. **Cost.** Refers to the cost of the setup.
vii. **Repeatability:** The perusing that fluctuates is more than once estimated under a similar situation.

4.2.2.2 CLASSIFICATION OF SENSORS

1. **Biosensors:** These depend on electrochemical innovation. They are utilized for food testing, clinical consideration gadget, water testing, and natural fighting operator recognition.
2. **Image Sensors:** These depend on the CMOS innovation. They are utilized in buyer gadgets, biometrics, traffic and security observation, and PC imaging.
3. **Motion Detectors:** These depend on infrared, ultrasonic, and microwave/radar innovation. They are utilized in video games and reproductions, light initiation, and security recognition.
4. **Accelerometers:** These depend on micro-electromechanical sensor innovation. They are utilized for understanding and observing, which incorporates pace producers and vehicle dynamic frameworks.

Some more affiliated terminologies have been given in Table 4.1.

TABLE 4.1 Presenting All Affiliated Terminologies of Proposed Framework

1.	Actuator	The actuator converts electrical signals (energy, usually carried by air, current, or liquid) into different forms of energy, such as movement or pressure. This is the opposite of what the sensor does, which captures the physical characteristics and converts them into electrical signals [4].
2.	Architecture	The basic organization of the system is embodied in its components, the relationship between them and the relationship with the environment, as well as the principles that guide its design and evolution.
3.	Connectivity	General term for connecting devices for data transfer and forwarding data transfer. Usually refers to the network connection, includes bridges, routers, switches, gateways, and backbones—can also refer to connecting a home or office to the internet or connecting a digital camera to a computer or printer.

TABLE 4.1 *(Continued)*

4. Device Discovery	Process of locating devices within the range of a network of a central device.
5. Embedded Computing/ Systems	Term for single-purpose computing, as opposed to general-purpose computing. The embedded computer system has a special purpose and only contains the software and hardware required to realize these purposes. On the internet of things, many systems are developed for specific purposes and work in conjunction with other systems.
6. Gateway	Network equipment or software runs on computers on the network, and they can communicate with other networks, even if these s use different protocols. Provide the ability to share information.
7. Internet of Everything	A term defined by Cisco Systems basically means applying the internet of things to everything, creating new features and intelligent processes in almost every area we can think of. Cisco calls it "people, processes, data, and things" connections. Network equipment or software runs on computers on the network and can communicate with other networks, even if they use different protocols. Provide the ability to share information
8. Microcontroller	A small computer on a single IC, containing a processor core, memory, and programable input/output peripherals. NOR flash memory or program memory in the form of OTP ROM is generally also contained on the chip, and generally a small amount of RAM. Compared to microprocessors used in personal computers or other general-purpose applications, microcontrollers are specifically designed for embedded applications. The microcontroller is used for automatic control products and equipment, such as automobile engine control systems, implantable medical equipment, remote control, office machines, electrical appliances, tools, and power toys. Compared to the design that uses an independent microprocessor, memory, and input/output device, by reducing the size and cost, the microcontroller can economically control even multiple devices and processes digitally. Mixed-signal microcontrollers are common and integrate analog components.
9. Pervasive Sensing	Perceive the ubiquitous ability of the device in activities or changes in conditions: It is considered indispensable on the internet of things and is generally applied to the system's perception of human activities.
10. Sensor Hub	A technology that connects sensor data and processes it. Data-processing job is being done.

4.3 LITERATURE REVIEW

A child-checking framework has been proposed and an upgraded clamor-dropping framework that can screen the infant and lessens sound contamination has been recommended. The principal capacity of the framework is to diminish the commotion that may upset the child by playing loosening-up melodies. This framework can likewise modify the room's light with a light sensor. Other E-infant support that can swing naturally when it discovers crying and quits swinging when the crying stops have also been proposed [2]. The speed of the swinging support can be controlled depending on the client's needs. It has a caution installed in the framework, which advises the client when two conditions happened. To begin with, the alert goes off when the sleeping pad is wet, showing that the bedding ought to be changed. Second, when the infant doesn't quit sobbing for a specific time, the caution makes the guardians aware of taking care of their child. It is just material when guardians are close to the support since it just uses a ringer caution, the sound of which may scare the infant. Guardians can't screen their infant when they are away from home, for instance, when at work or when heading out to different spots. A comparable programed child observing framework was proposed in Ref. [3]. The creators built up a low-spending framework that swings the support when the crying sound is identified, and the support stops when the infant quits crying. The inherent alert goes off under both of the accompanying conditions: the sleeping pad is wet, or the infant doesn't quit crying after a specific period. A camcorder is put over the support to screen the child. Be that as it may, the guardians can just get the warning through SMS and can't control the framework. In this manner, the proposed framework in the current examination is further developed, in light of the fact that it uses an IoT application to screen and control the created shrewd support continuously anyplace and whenever. An Arduino-based resonant support structured with infant cries recognition was proposed [4]. A metal ball configuration is embraced to diminish framework damping and permits the support to swing uninhibitedly even without power. Along this, a fitting sensor is intended to distinguish the swinging status or edge. The authors guarantee that their framework is vitality-sparing and permits guardians to record newborn child cries because of yearning or agony on an SD card. In any case, such a nearby control arrangement is wrong when guardians are found marginally a long way from their children, since it doesn't permit refreshing of the information in the IoT server or controlling the support remotely. Structured a framework for infant checking dependent on Raspberry Pi and Pi cameras [5]. The structured framework can detect the movement and crying state of the infant. They utilized condenser MIC to detect the crying condition and a PIR movement sensor to recognize the child's development with the assistance of a Pi camera. The camera is turned on just when the condenser MIC identifies a sound and imparts a sign to Raspberry Pi. Nonetheless, the yield of this framework is just accessible on-screen show; along with this, the guardians can just view the information on a set number of gadgets inside a fixed region. In Ref. [6], the authors proposed a framework that can screen the heartbeat rate and internal heat level of the individual. Committed sensors are put alongside Raspberry Pi and IoT to screen the well-being condition and store the acquired information. SYHS2XX arrangement is utilized to check the glow around the child and moisture of the hatchery as proposed by Joshi, Kamat, & Gaikwad [9]. In their model, two sorts of tests are utilized implied for skin-temperature test and air temperature test, which are used to screen the temperature around the newborn child and the clamminess of the incubation center. A newborn child-checking framework utilizing numerous sensors was created [10]. This chapter receives remote correspondence mode, for example, Bluetooth and RS232 sequential correspondence modem for structuring a completely incorporated proficient well-being observing framework for babies, utilizing wearable sensor frameworks for estimating ECG, temperature, and CO_2 level around the baby's support or bassinet. The framework utilizing J2ME innovation is utilized in various fields to create programs for versatile remote data gadgets, for example, PDAs and individual advanced associates (PDAs). The above said innovation was utilized in infant observation, utilizing Video Surveillance System Using Motion Detection Method for live spilling of the babies at whatever point the movement was experienced utilizing cross connection [11]. To conquer the hindrances of costly movement indicators "A minimal effort, portable based checking, and warning framework" was created to screen the newborn child and indirectly concern the mother on kid status [12]. This investigation incorporates perpetual estimating of the glow, pulse, and development and sends it to a cloud where the data is dealt with. By using an Arduino UNO as the inside processor an IoT bunk was structured by Chun-Tang Chao, Chia-Wei Wang, Juing-Shian Chiou, & Chi-Jo Wang [4]. The fundamental target of their work is to diminish the vitality by embracing the sensors which are planned to distinguish the faltering state. The principal safe and naturally shaking support was designed by Marie R. Harper. The spring-stacked engine was utilized to give oscillatory movement to shake the bunk as it would be shaken by the mother [14]. Further, the customary lodging is appended with the electronic gadget and is electronically activated for shaking. It comprises affectability control with the goal that the child's crying voice, distinguished by the receiver, can incite shaking activity for some time using a clock [15]. Later a gadget is designed to recognize the infant's crying voice. The sound sign is intensified with an enhancer, and further, the heartbeat signal is created by a heartbeat generator circuit. This heartbeat signal contributes to a sign acknowledgment circuit which demonstrates infant cry discovery as a yield [16]. A programmed infant rocker having commotion sensor to identify infant cry is proposed where the preamplifier intensifies the input sound sign. Arduino ATmega328P microcontroller is used to control dc engine for shaking meanwhile to engage the child while shaking, bright LED lights are used [17]. A calculation is proposed by Yang Hu to change the bunk's influencing degree by signs of sensors. To quantify child status, three weight sensors are utilized in sensor organization. Binu et al. have proposed a system that uses Hadoop and the C4.5 algorithm for predicting the disorders using the collected data. It monitors the baby and gives an update on the health and mental status of the children. Including more health-based sensors in the system will help in the health monitoring and guide medicinal care in case of any abnormality [42]. Sagar et al. Proposed a method that uses the IoT, Amazon web services, and smart baby cribs, and provides parents with a smart system to help these parents monitor and comfort their babies. The crib system is a device used in place of the caregiver to comfort children by playing music and talking to them. The current job reduces labor, especially the pressure from the mother during working hours. The entire mechanism is mobile and can be easily moved from room to room [43]. Lai et al. proposed a framework suitable for childcare. Deep learning technology was used at times. The device monitors the baby's body position, body temperature, posture, and other conditions through deep learning to help parents understand the baby's condition. It may contain more important features [13]. Srinivas et al. in his development included a serial camera and a LinkIt ONE board for guardians to locate the exact location of the child and provide health information to the child through messages

[39]. Simon et al. used the Raspberry Pi B + module to control the hardware system. In the development, MIC has been used to detect the baby's crying, and the built-in PIR motion sensor is used to detect the baby's movement [40].

4.4 METHODOLOGY

4.4.1 RESEARCH DESIGNING

Right from observation to the final development and deployment of the proposed research, the various steps that it involves, have been shown in Figure 4.3.

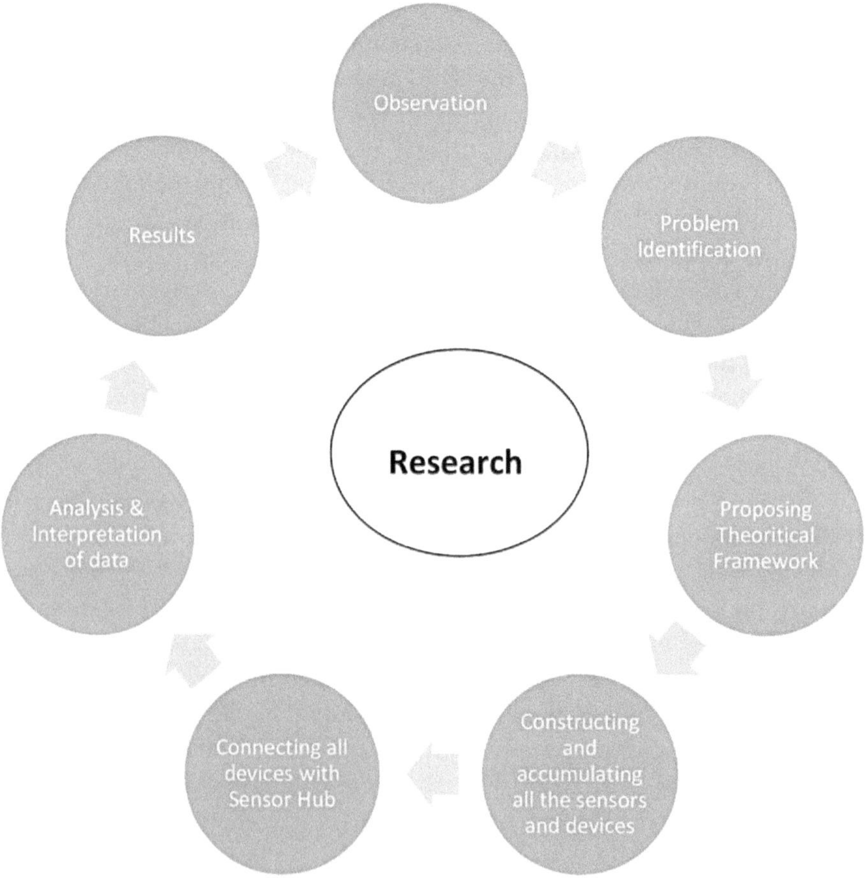

FIGURE 4.3 The research design for proposed research.

4.4.2 UTILIZATION

The proposed framework can be utilized by moms to deal with their children in their habitation. This framework will caution the mother or the person looking after the baby through SMS. In addition, when the mother gets the SMS, she can see the infant through online observation and can take concerned actions. On the other hand, if the mom is exceptionally a long way from the infant, she can demand any neighbor or any person who is a close relative to proceed to deal with the child. Through this framework, the mother will get an SMS at whatever point the infant is crying, moving, or peeing. Along with these features, this framework holds an additional feature of security; if any interloper binds to hurt the child, the camera will catch the picture of that individual and will alarm the mother or the caretaker of the baby.

4.4.3 COMPONENT SELECTION

The various gadgets that we need for our artificial mother monitoring system of infant baby include:

1. **Raspberry Pi 3 Model B+:** The Raspberry Pi 3 Model B+ has dual-band 2.4 GHz and 5 GHz wireless LAN, Bluetooth 4.2/ BLE, faster Ethernet, and PoE functionality through a separate Power over Ethernet HAT, in addition to a 64-bit four core CPU running at 1.4 GHz. Amazon is already taking preorders for the Raspberry Pi 3 Model B+. The board offers dual-band wireless LAN compliance certification as well as modular compliance certification, which enables it to be incorporated into end devices with significantly less wireless LAN compliance testing, lowering both cost and time to market. The Raspberry Pi 3 Model B+ has the same physical footprint as the Raspberry Pi 2 Model B and Raspberry Pi 3 Model B, and it's also backward compatible with the Raspberry Pi 3 Model B.
2. **PIR Motion Sensor:** A passive infrared (PIR) sensor is an electronic sensor that detects, and records infrared light generated by objects within its field of view (FoV). They are most often encountered in PIR-based motion detectors. A PIR sensor is a kind of sensor used in security alarms and controlled lighting systems.

On a technological level, PIR technology is made up of a pyroelectric sensor that can detect variable amounts of infrared light. Example: Everything emits varying degrees of radiation, with the strength rising in direct proportion to the item's temperature. In reality, the motion detector is split into two portions since we're more concerned with motion change than with IR level. The output will swing between high and low if one side of the gadget detects different infrared radiation than the other.

3. **Sound Sensor Module Raspberry Pi:** Sound sensor act like a microphone that detects sound signals. It takes up sound signals and converts them into electrical signals.
4. **High-Sensitivity Water Sensor:** The Water Sensor module is a piece of the Grove framework. It demonstrates whether the sensor is dry, moist, or totally drenched in water by estimating conductivity. The sensor follows have a powerless draw-up resistor of 1 Mω. The resistor will pull the sensor follow esteem high until a drop of water shorts the sensor follow to the grounded follow. In all honesty, this circuit will work with the computerized I/O pins, or you can utilize it with the simple pins to recognize the measure of water-instigated contact between the grounded and sensor follows.
5. **Temperature Sensor:** The DS18B20 temperature sensor is ideal for ventures like climate stations, home robotization frameworks, or monitoring. Hardly any sensors are this simple to set up on the Raspberry Pi. They're a similar size as a transistor and utilize just one wire for the information signal. They're likewise incredibly precise and take estimations rapidly. The main segment you need is a 4.7K Ohm or 10K Ohm resistor.
6. **Speakers:** Your Raspberry Pi robot will have a voice and sound due to the Speaker for Raspberry Pi robot. This speaker connects to the Raspberry Pi's aux jack and can play music, speech, or warnings. The sound produced by the Raspberry Pi Speaker is rich and resonant. It elegantly has its capacity that may be blamed on the USB port.
7. **Pi Camera:** In April 2016, the initial Camera Module was replaced with the Raspberry Pi Camera Module v2. A Sony IMX219

8-megapixel sensor powers the v2 Camera Module (contrasted with the 5-megapixel OmniVision OV5647 sensor of the first camera). The Camera Module, like still photographs, may be used to capture high-quality video. It's simple to use for beginners, but it has a lot to offer advanced consumers who want to expand their knowledge. There are several examples of people who use it for time-passing, slow-motion, and other video capabilities on the internet. To produce impressions, you may also use the libraries we've gathered using the camera.

4.4.4 WORKING

This checking framework ought to be physically turned on when the mother wishes to disregard the infant sleeping in the house. In this framework, there are four tactile gadgets, for example, a PIR sensor to screen the physical development of the infant, a dampness sensor to detect whether the child has peed, at that point a clamor sensor to identify whether the infant is crying and a camera for recording the visuals. Every one of these sensors is associated with an Arduino chip, which follows a calculation to process the tactile information and give the yield. So, at whatever point the infant moves from its position, the GSM module sends an instant message to the mother and afterward, she can watch the child through online observation and can make a move. Essentially if the infant awakens and cries or discharges, the sensors will identify and the mother will be promptly informed through instant messages, with the goal that the mother can ask help from someone (like a neighbor) to deal with the child. Or, more than likely, she has the alternative to play pleasant music through the speakers to make the straight nod off once more. If there should be an occurrence of any danger, as if any interloper attempts to hurt the infant, the PIR sensor will detect it, and the camera will consequently begin recording the visuals. At that point, this account can be utilized as proof to grasp the crook.

4.4.5 FLOWCHART

Switch on the device and check whether it's working or not else detect an error in a circuit. If it's working, keep the device 2 cm away from the baby. Keep the moisture sensor in a diaper. Also, keep the camera at an appropriate view angle from which the baby is clearly visible. Now connect the camera to your mobile phone for online surveillance over the IP. Else if the motion is detected, switch on the camera and record the video. If moisture and cry of the baby are detected, inform the caretaker or mom via SMS. Mom or caretaker will come and will take the concerned action. Since the baby is awake now, one can switch off the device. Switch on again when the baby is sleeping and when you want to leave that room for other work (Figure 4.4).

4.4.6 CIRCUIT DESIGNING

The black wire represents the GND terminal of the circuit. The Red wire shows the 5 V power supply (also called VCC) for the circuit. The remaining wires are used for collecting data from the sensors or sending signals to the relay module for controlling the appliance. The camera is directly connected to the PI directly using the inbuilt connector (Figure 4.5).

4.5 RESULTS

After the deployment of the proposed idea, various phases of analysis of data collection and preprocessing, the final output, according to the expectations, were recorded with the accuracy of 92% of systems acting in coordination. The environment of the artificial mother monitoring system has been tested in 5 different environments (motion, moisture, noise, temperature, and altogether combination). The reactive time was in coordination with 92% of accuracy. Meanwhile, the data collection and preprocessing of data has achieved 100% accuracy.

4.6 APPLICATION OF PROPOSED RESEARCH

In enormous MNCs, there is a large number of ladies laborers who take care of the babies of the moms who are working in that company where moms work with a sort of tension of baby care in their heart. In such cases, this framework can be deployed. It will provide the moms with more

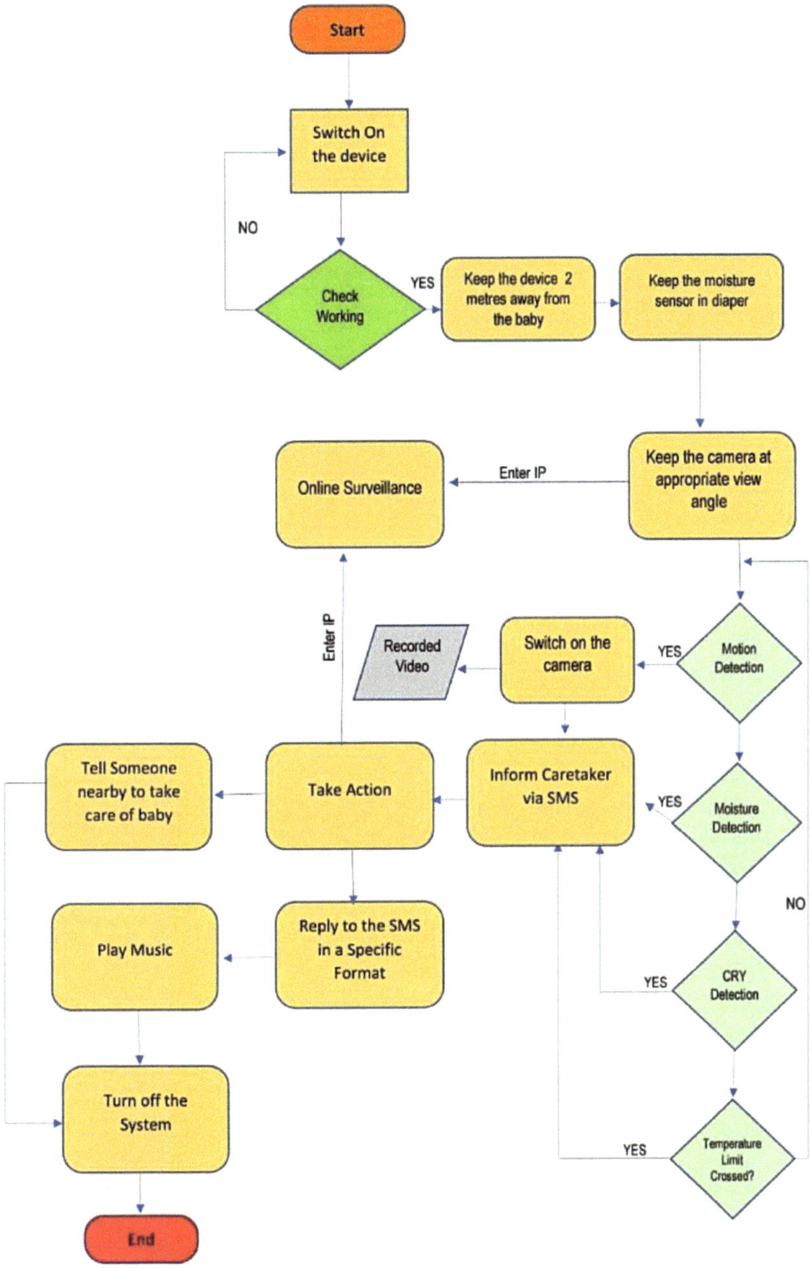

FIGURE 4.4 Artificial mother monitoring system for a sleeping infant.

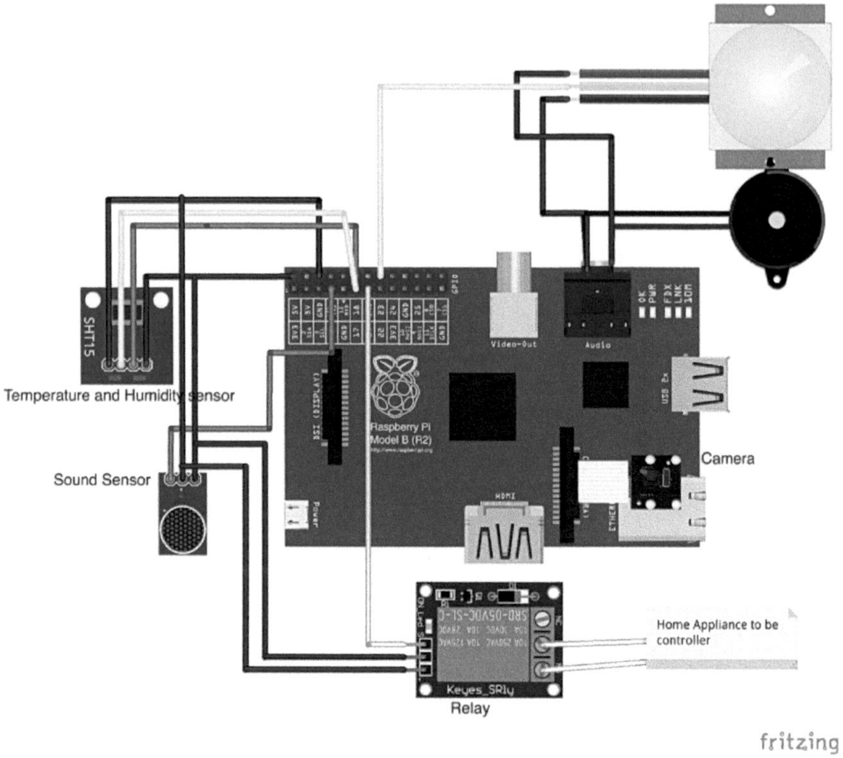

FIGURE 4.5 Circuit design of the system.

Source: The circuit is designed using Fritzing software [19].

security satisfaction and rather they can concentrate more on their work. In each clinic, there is an extraordinary ward where infants are kept. In this ward, this framework can be introduced to screen the infants and deal with them. It may be deployed in an in-house environment; it will assist moms who are alone at home and have to deal with various household chores and babies at the same time.

4.7 CONCLUSION

In this framework, a shrewd infant observation is proposed dependent on IoT to achieve the distantly checking of different prosperity boundaries and telling the people who take care of the infant. The proposed system is intended for a progressed computerized child-rearing condition and remembering that – to an extraordinary degree inappropriate – security concerns may put some off. Additionally, the system can decrease the correspondence hole between the mother and the newborn child. This work shows the undertaking to shield the prosperity of children and gives comfort and doubt that "everything is great and acceptable" to the person who is watching a baby.

4.8 FUTURE SCOPE

The proposed framework can be taken to an advanced version of components and hence can be made interoperable. A proper coordinating application can be developed. More advanced algorithms of machine learning can be used in order to save battery, power, and storage.

KEYWORDS

- **artificial intelligence**
- **Internet of Things**
- **machine learning**
- **monitoring system**
- **security**
- **sensor nodes**
- **smart parenting**

REFERENCES

1. Brangui, S., Kihal, M. E., & Salih-Alj, Y., (2015). An enhanced noise canceling system for a comprehensive monitoring and control of baby environments. In: *2015 International Conference on Electrical and Information Technologies (ICEIT)* (pp. 404–409).
2. Goyal, M., & Kumar, D., (2013). Automatic E-Baby Cradle Swing based on Baby Cry. *International Journal of Computer Applications 71*(21), 39–43.

3. Palaskar, R., Pandey, S., Telang, A., Wagh, A., & Kagalkar, R., (2015). An automatic monitoring and swing of the baby cradle for infant care. *International Journal of Advanced Research in Computer and Communication Engineering, 4*(12), 187–189.
4. Chao, C. T., Wang, C. W., Chiou, J. S., & Wang, C. J., (2015). An Arduino-based resonant cradle design with infant cries recognition. *Sensors, 15*(8), 18934–18949.
5. Symon, A. F., Hassan, N., Rashid, H., Ahmed, I. U., & Reza, S. M. T., (2017). Design and development of a smart baby monitoring system based on Raspberry Pi and Pi camera. In: *2017 4 International Conference on Advances in Electrical Engineering (ICAEE)* (pp. 117–122).
6. Kaur, A., & Jasuja, A., (2017). Health monitoring based on IoT using Raspberry Pi. In: 2017 *International Conference on Computing, Communication and Automation (ICCCA)* (pp. 1335–1340). IEEE.
7. Andry, G. F., Duarte, F., & Syrgio, B., (2016). A smart wearable system for sudden infant death syndrome monitoring, industrial technology (ICIT). In: *2016 IEEE International Conference*. IEEE, ISBN: 978-1-4673-8075.
8. Marti, P., & Andreoni, G., (2008). Pervasive technology in neonatal intensive care unit: A prototype for newborns unobtrusive monitoring. In: *30 Annual International IEEE EMBS Conference Vancouver.* British Columbia, Canada, 978-1-4244-1815-2/08/,1292-1295.
9. Joshi, N. S., Kamat, R. K., & Gaikwad, P. K., (2013). Development of wireless monitoring system for neonatal intensive care unit. *International Journal of Advanced Computer Research, 3*(11).
10. Sakshi, G., Zarnain, M. K., Rupali, S., & Chougule, P. A., (2016). Infant monitoring system using multiple sensors. *International Journal of Research in Engineering and Technology, 5*(5).
11. Pramod, G., Shailesh, B., & Prasad, W., (2013). Wireless automated video surveillance system using motion detection method. *International Journal of Engineering Research and Applications (IJERA), 3*(2), 863–865. ISSN: 2248- 9622.
12. Owen, N. N. F., Nii, H., Adrian, D. C., Ajith, P. M., Gopalakrishnakone, P., & Zubair, A., (2011). *Low-Cost Infant Monitoring and Communication System.* IEEE International Conference Publication, Science, and Engineering Research.
13. Chinlun, L., & Lunjyh, J., (2018). an intelligent baby care system based on IoT and deep learning techniques. In: *The Year 2018, International Scholarly & Scientific Research & Innovation.*
14. Marie, R. H., La Mirada, & Maxine, R. B., (1973). *Automatically rocking baby Crdle.* US3769641.
15. Gim, W., (1976). *Automatic Bay Crib Rocker*. US3952343.
16. Chau-Kai-Hsieh, Chiung, L., & Taiwan, (1997). *Baby Cry Recognizer*. US5668780.
17. Steven, B., Richard, L., & Natalia, L., (2011). *Rock Me Baby: The Automatic Baby Rocker.* Project for, San Jose State University, Department of Mechanical and Aerospace Engineering.
18. https://www.theguardian.com/technology/2016/oct/26/ddos-attack-dyn-mirai-botnet (accessed on 21 December 2022).
19. https://fritzing.org/ (accessed on 21 December 2022).
20. Manh, K., (2015). *Sensor Communication in Smart Cities and Regions: An Efficient IoT-Based Remote Health Monitoring System.* Master of Science, computer science

and engineering, Luleå University of Technology Department of Computer science, Electrical and Space Engineering, Sweden.
21. Taneja, R., (2018). *Protection of Personal Information Used by IoT Devices in Health Care*. Masters in Computer Science at Flinders University, Adelaide, Australia.
22. Chopra, et al., (2019). *IoT-Based Patient Monitoring System, Bachelor of Technology in Electronics and Communication Engineering*. Jaypee University of Information Technology, Waknaghat.
23. Marx, G. F., (1979). *Clinical Management of Mother and Newborn.* ISBN 978-1-4612-6173-5: Springer.
24. Sarojini, et al., (2007). Challenges faced by new mothers in the early postpartum period: An analysis of comment data from the 2000 pregnancy risk assessment monitoring system (PRAMS) survey. *Matern. Child Health J., 11*, 549–558. Springer Science+Business Media, LLC 2007.
25. Pecchia, L., et al., (Eds.) (2014). IWAAL 2014. Garment Design for an Ambulatory Pregnancy Monitoring System, LNCS 8868, pp. 219–227, 2014. © Springer International Publishing Switzerland 2014.
26. Zhang, et al., (2018). Medical long-distance monitoring system based on internet of things, EURASIP. *Journal on Wireless Communications and Networking, 2018*(1), 1–8, 176. https://doi.org/10.1186/s13638-018-1178-2.
27. Khan, T., (2021). An intelligent baby monitor with automatic sleeping posture detection and notification. *AI, 2*, 290–306. https://doi.org/10.3390/ai2020018.
28. Sarhaddi, F., Azimi, I., Labbaf, S., Niela-Vilén, H., Dutt, N., Axelin, A., Liljeberg, P., & Rahmani, A., (2021). Long-term IoT-based maternal monitoring: System design and evaluation. *Sensors, 21*, 2281. https://doi.org/10.3390/s21072281.
29. Fazal, N., Webb, A., Bangoura, J., & El Nasharty, M., (2020). Telehealth: Improving maternity services by modern technology. *BMJ Open Qual., 9*, e000895.
30. Sharma, S., Sidhu, H., & Kaur, S., (2016). Analytical study of intrauterine fetal death cases and associated maternal conditions. *Int. J. Appl. Basic Med. Res., 6*, 11.
31. Souri, et al., (2020). A new machine learning-based healthcare monitoring model for student's condition diagnosis in internet of things environment. Springer-Verlag GmbH Germany, part of springer nature. *Soft Computing.* https://doi.org/10.1007/s00500-020-05003-6.
32. Malik, H., Fatema, N., & Alzubi, J. A., (2021). *AI and Machine Learning Paradigms for Health Monitoring System*. Springer, ISBN 978-981-334-412-9.
33. Krzysztof, J., (2019). *Free Baby Monitoring System with ML for Safe Activity Transmission.* https://www.netguru.com/blog/free-baby-monitoring-system-with-ml-for-safe-activity-transmission (accessed on 21 December 2022).
34. Vemuri, et al., (2020). Artificial intelligence and internet of medical things-based health-care system for real-time maternal stress – strategies to reduce maternal mortality rate. *Drug Invention Today, 13*(6). ISSN: 0975-7619.
35. https://dataconomy.com/2021/03/ai-maternal-and-fetal-medicine-new-model-mother-child-care (accessed on 21 December 2022).
36. Niela-Vilen, H., Rahmani, A., Liljeberg, P., et al., (2020). Being 'a google mom' or securely monitored at home: Perceptions of remote monitoring in maternity care. *J. Adv. Nurs.,76*, 243–252. 10.1111/jan.14223.

37. Dipali, B., Neha, S., & Dnyaneshwar, K., (2019). Smart and secure IoT-based child monitoring system. *International Research Journal of Engineering and Technology (IRJET)* (Vol. 06, No. 11). E-ISSN: 2395-0056.
38. Sohini, R., & Uma, B., (2015). *Smart Mom: An Architecture to Monitor Children at Home*. Conference Paper, WCI '15, Kochi, India © 2015 ACM. doi: 10.1145/2791405.2791540.
39. Nandini, P. M., Murugan, S., Srinivas, K. N. H., Sarveswararao, T. D. S., & Kusuma, K. E., (2019). *Smart IoT Device for Child Safety and Tracking*. Published By: Blue eyes intelligence engineering retrieval number & sciences.
40. Aslam, F. S., Nazia, H., Humayun, R., Iftekhar, U. A., & Taslim, R. S. M., (2017). *Design and Development of a Smart Baby Monitoring System Based on Raspberry Pi and Pi Camera.* IEEE:
41. https://www.thequint.com/news/india/working-mothers-in-india-report-ashoka-university (accessed on 21 December 2022).
42. Binu, P. K., Akhil, V., & Vinay, M., (2017). *Smart and Secure IoT-based Child Behavior and Health Monitoring System using Hadoop.* IEEE.
43. Sagar, S. B., & Nilkanth, B. C., (2018). IoT-based healthy baby cradle system. *IJRIEE.*

CHAPTER 5

A Review of Smart Intelligent IoT Network: Technologies and Real-Time Applications

GAJENDRA KUMAR AHIRWAR, RATISH AGARWAL, and ANJANA PANDEY

[1]*Research Scholar, University Institute of Technology RGPV, Bhopal, Madhya Pradesh, India*

[2]*Associate Professor, University Institute of Technology RGPV, Bhopal, Madhya Pradesh, India*

ABSTRACT

Man constantly tries to make every task simple and convenient in his busiest life, as a result of which he makes new discoveries and research every day. In this computer era, everything from daily work to machinery work is being done very easily. In this sequence, another new discovery was made to give a new dimension to physical objects, in which all the things were attempted to be done smoothly by connecting all the physical objects through the internet. This new technology was called the internet of things (IoT). This technology gave digitalization form to the social and economic life of man, where every object (living or non-living) communicates with each other with the help of a computer network. In the last few years, computer technologies are developed rapidly, a result of which IoT is used to make physical entities smart and to process intelligent behavior with the help of an internet connection. In this chapter, we present an overview of IoT networks with their technologies, intelligent applications, and challenging task.

Intelligent Sensor Node-Based Systems: Applications in Engineering and Science, Anamika Ahirwar, Piyush Kumar Shukla, Prashant Kumar Shukla, and Ruby Bhatt (Eds.)

5.1 INTRODUCTION

Kevin Ashton was the first person who define the term IoT or internet of things in 1999. He used this term to describe supply chain management, where the real-time entities can be managed through the computer system via internet connectivity.

To simply describe the IoT, we can say that it is a system that connects the different objects or items of the physical world as a network where these objects are connected to each other with the help of different interfaces, protocols, and sensors via the internet connection. This internet connection may be wired or wireless. The base of IoT is the internet; it is not possible to create an IoT network without internet connectivity.

In human life, IoT provides smartness to everything in an indoor or outdoor environment as the devices have been able to send and receive data from one to another in seconds. In other words, we can say that IoT is a kind of family of devices, sensors, hardware software, and various living and non-living things that can collect and exchange information through the Internet. This can be easily understood by an example, suppose we are going on a dark road and the streetlights on that road are managed by the IoT network, then as soon as we move on that road, we will be sensed by the streetlight, and it will turn on automatically. Thus, as we move along the road, the sensor information sent by one streetlight will be passed on to another streetlight and thus the lights will turn on and off as we go forward. This process will be done with the help of IoT or IoT networks (Figure 5.1).

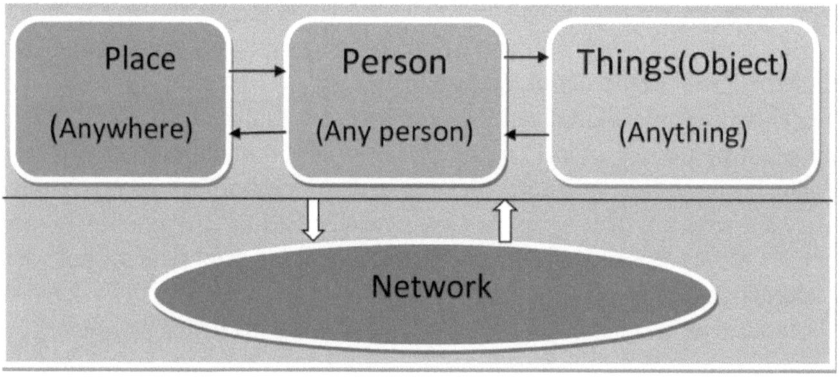

FIGURE 5.1 Internet of things (IoT) network.

5.2 ELEMENTS OF IOT

The internet of things (IoT) consists of multiple components that are necessary to build up a strong device-to-device IoT network. A lot of basic elements that are grouped in an IoT network are as follows.

5.2.1 SENSORS

Sensors are a type of electrical device that takes the input of physical data like heat, light, moisture, pressure, speed, etc., from the environment and converts them into signals as energy through electronic devices. These signals are measured electrically.

Sensors are used to detect and measure. Sensors are basically categorized into two main categories.

- Analog/digital sensor; and
- Active/passive sensor.

The analog sensor works on position, velocity, etc., and the digital sensor generates binary signals or binary data with respect to their input value. Active sensors are sensors that have their own source of energy, while passive sensors work by the energy emitted by the sun. There are various types of sensors used in IoT, for example, temperature sensors, IR sensors, proximity sensors, pressure sensors, moisture sensors, level sensors, optical sensors, etc. (Table 5.1).

TABLE 5.1 Analysis of Different Types of Sensor Used in IOT

Type of Sensor	Input Data	Applications	Example
Temperature Sensor	Heat	Refrigerators, ACs	Thermocouples, Thermistors, DHT11
IR Sensor	Energy (Heat)	Home appliances, infrared vision	MLX90614, ISL29021
Pressure Sensor	Force	Mobile phones, fluid flow, gps navigations	BMP180, BP10, KELLER
Proximity Sensor	Motion	Retailing, parking space, etc.	Si1102, Si114X
Smoke Sensor	Gas	Gas detectors	MQ-2
Level sensor	Liquid or material level	Manufacturing, automotive industries, level management	XMP800 series

5.2.2 ACTUATORS

These devices take the electrical signals as an input and convert them to their corresponding physical quantities. An actuator is a type of transducer which works as opposed to sensor devices. It is also responsible to control and manage the system devices. There are basically two types of actuators.

- binary actuator; and
- continuous actuator.

Examples of actuators: servo motors, DC motors, relays, etc.

5.2.3 GATEWAY

It is a middle layer between the devices and the network in an IoT system. Gateway takes the data from sensors or devices and sends it to the network for further processing. It is also responsible for device-to-device and device-to-cloud communication. The gateway can be a software or hardware device.

5.2.4 PROTOCOLS

Protocols are the standards or set of rules to transmit data (information) between various devices over the same or different network. IoT protocols are very important to exchange data between sensors and the cloud (network) (Table 5.2).

5.2.5 END USER/DEVICES

End users or devices are used to control the system which is deploying in the IoT network. These devices interact with the other devices and collect the data which is transmitted to the other end.

5.3 IOT TECHNOLOGIES

The internet of things (IoT) provides communication between devices using a variety of technologies. All these technologies are very important for building an IoT network. Some of the technologies used in IoT are as follows.

TABLE 5.2 Different Types of IOT Technologies and Their Protocols

Protocol	Topology used	Applications	Functions	Drawback
Bluetooth	PAN	Smartphones	Short range communication	Limited distance, one to one device
Zigbee	MESH	Home automation	Low power communication	Limited speed, limited distance
LoRAWAN	STAR	Public network	Large scale bi-directional communication	Not suitable for real time applications
6LOWPAN	MESH	Home , office	Lightweight IP based communication	Limited processing
CARP	MESH	Under water communication	Network initialization , data forwarding	Not support existing data collection
MQTT	STAR	Monitoring sensor nodes	Provide embedded connectivity, M2M communication	Device overhead
CoAP	Multihop	Home automation, health care, transportation	Wide area geographical coverage	Unreliability of data transmission

5.3.1 RFID

RFID stands for radio frequency identification. It is a wireless communication technology that uses radio frequency or electromagnetic waves. This technology automatically identifies the object which has an RFID tag and collects data from them. In this technology, there are two main components used for communication, tags, and reader.

The RFID tag is a small microchip that stores the data. It is activated by the signals sent by the RFID reader. There are two types of RFID tags that are active and passive. Power is required for the active RFID tag while the passive tag receives power from the electromagnetic waves of the reader. The RFID reader consists of a scanning antenna and transceiver that transmits radio waves and receives the RFID tag signals.

RFID tags are different from barcodes in that the data keeps updating while the data stored in the barcodes cannot be changed. Also, RFID does not require line of sight; it reads automatic data when it comes in range.

It is mainly used in commercial, industrial, and inventory applications (Figure 5.2).

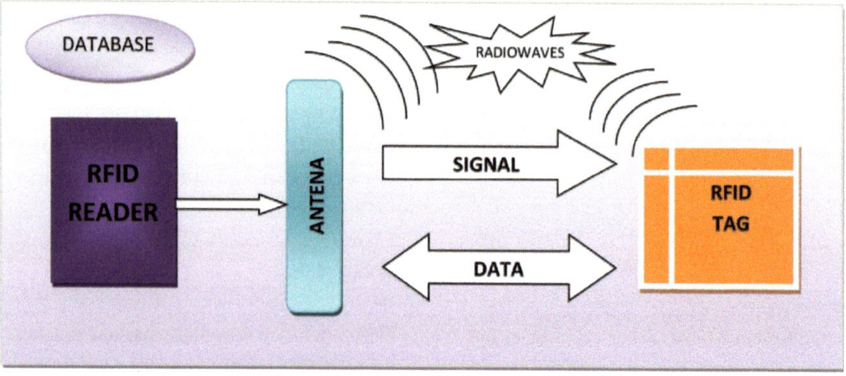

FIGURE 5.2 RFID working procedure.

5.3.2 NFC

Near-field communication (NFC) is a wireless communication technique. It has short-range communication between the devices. There is no need to pair the devices like Bluetooth. If NFC-enabled devices are in a range, they can be directly communicated to other NFC devices or tags. The NFC device generates the magnetic field. This magnetic field propagates the electromagnetic (radio) signals which are received near NFC tags or devices. The radio signals are decoded by the end device for getting the essential information. There are basically two types of NFC devices.

If a device has its own electromagnetic fields, then it is called an active NFC device while the device that needs external power is called a passive NFC device.

5.3.3 WI-FI

Wireless fidelity or WI-FI is a wireless technology, which uses electromagnetic or radio waves for communication. Radio signals, antenna, and wireless routers are basically the three main components used in a wireless network. In 1999 IEEE define a standard as 802.11b for wireless fidelity. It is majorly used in wireless local area networks which already built-in laptop computers, smartphones, smartwatches, and some other types of equipment. It is used as a WIFI hotspot. This standard operates on 2.4

GHz ISM frequency band. In this technology, the carrier senses multiple accesses with collision avoidance (CSMA/CA) are used when transmitting the data packets. Some other IEEE 802.11 standards are summarized here which are used by Wi-Fi (Table 5.3).

TABLE 5.3 Summary of Different IEEE Standards for Wireless Network

Standard	**Release date**	**Frequency**	**Typical data rate**	**Maximum data rate**	**Indoor area range**	**Outdoor area range**
802.11a	1999	5.1–5.8 GHz	25mbps	54 mbps	25 M	75 M
802.11b	1999	2.4–2.5 GHz	5.5 mbps	11 mbps	35 M	100 M
802.11g	2003	2.4–2.5 GHz	25 mbps	54 mbps	25 M	75 M
802.11n	2007	2.4–5 GHz	200 mbps	500 mbps	50 M	

5.3.4 BLUETOOTH

Bluetooth's name comes from the Denmark king Harald Blatand. Bluetooth technology is developed in 1994 by Erricson. It is a short-range wireless communication technology that creates a personal area network (PAN) or piconet. The maximum range for Bluetooth is 10 to 15 meters. Bluetooth is a one-to-one communication standard. When one Bluetooth-enabled device is active and other Bluetooth-enabled devices are in its range then firstly, they need to pare and after pairing the devices can be communicated with each other. There is a limitation in that only one device can be able to transfer data at that time. IoT works with different Bluetooth versions. The current version is Bluetooth 5.0 which was released in 2017 (Table 5.4).

TABLE 5.4 Comparison Between Different Wireless Technologies Used in IOT

Characteristics	**RFID**	**NFC**	**WiFi**	**Bluetooth**
Acronym	Radio frequency identification	Near field communication	Wireless fidelity	Bluetooth
Frequency	LH-UHF	13.5 MHz	2.5–5 GHz	2.4 GHz
Data rate	Always vary according to frequency	100–450 Kbps	145 mbps	20–25mbps
Coverage area	2–10 meters	5–10 cm	100–500 meters	10–100 meters
Communication mode	Unidirection	Bi-direction	Bi-direction	Bi-direction

5.4 APPLICATIONS OF IOT

The internet of things (IoT) has many applications which work in different areas. These applications have specific functionality which provides a smart and easy workflow to people. Some applications and specific areas are listed below:

- Educational;
- Healthcare;
- Home appliances;
- Agriculture;
- Transportation;
- Consumer services;
- Engineering and manufacturing.

5.5 RELATED WORK

In this section, we mentioned different literature works which are done in the last 10 years. There are various applications where IoT does better and creates a smart application for better and easy human lives (Table 5.5).

5.6 EXAMPLE OF INNOVATIVE IOT SMART APPLICATIONS

5.6.1 SMART GARBAGE SEPARATOR

To make cities smart and developed, many activities are being done on different dimensions. The most important of these is the systematic disposal of waste. In this era of the development of cities, it has become a challenge to organize the wet, dry, and chemical waste generated by factories and other means in a planned manner. To solve this problem, we suggest a smart dustbin through IoT, which will enable proper disposal of waste through IoT devices and sensors (Figure 5.3).

The trash will be placed in the center of the moisture-detecting doors, which contain the opening on the top. The open/close doors are driven using servo motors. The function of the moisture-detecting doors is to detect the dry/wet waste and take command from a microcontroller and

TABLE 5.5 Related Work on Internet of Things (IoT) Applications

Authors	Publication year	Application type	Objective	Technology used	Methodology	Functional outcomes
Zhai et al. [1]	2017	Smart Cities	Crowd sensing task assignment	Cognitive radio network	• PSO-CTA algorithm • Principles of crowdsensing	Assigning wireless spectrum sensing to mobile intelligent nodes
Beltran et al. [2]	2017	Smart Cities	User privacy in smart cities	IOT-ARM	• SMARTIE integrating IOT platform	Empower citizens to take control of their privacy policies and devices
Imran et al. [3]	2018	Smart Cities	Indoor positioning system	WLAN	• KLFDA positioning algorithm • Features extraction of RSS	Efficiently position estimation
Muhammad Alam et al. [4]	2018	Smart Cities	Smart parking system	LORA	• Distributed architecture • Ad hoc computer vision • Sensor based parking system	Vision based intelligent parking
Abd el Hamid et al. [5]	2018	Smart Cities	Road safety in smart cities	Exploiting advanced in IOT	• Design dynamic assessment • Hidden Markov modeling	Robust and dynamic road safety assessment
Francesc Alias et al. [6]	2019	Smart Cities	Noise monitoring	Wireless acoustic sensor network	• Design ad-hoc acoustic sensors • Pervasive deployment of noise monitoring network	Monitor specific noise sources in urban network
Daniel Minoli et al. [7]	2019	Smart Cities	Issues related to 5G based IOT applications in smart cities	NB-IOT and LTE-M	• Use cases described by various industries	–

TABLE 5.5 *(Continued)*

Authors	Publication year	Application type	Objective	Technology used	Methodology	Functional outcomes
Yongchun Cao et al. [8]	2020	Smart Cities	Sustainable development of smart cities	Cloud computing	• Financial cloud intelligent • Empirical analysis	Integration of smart city and financial cloud
Aggeliki Vlachostergian et al. [9]	2016	Smart home automation	User adaptive & context aware smart home	Pervasive and semantic technology	• Semantic interoperability • Knowledge management • Context model representation • User modeling apprach	User modeling and context awareness smart home
Noel Nuo Wi Tay et al. [10]	2016	Smart home automation	Smart home automation and optimization	SOA, Pervasive computing	• Bacterial memetic algorithm • Weighted constraint satisfication problem • Branch & bound theory	Automatic planning for Smart home appliances
Salihu Oladimeji Aliyu et al. [11]	2017	Smart home automation	Low cost GSM Bluetooth home automation	Bluetooth and GSM	• Develop an android application • Feedback system using sensors • Security using PIR sensor	Easy to use and control Smart home system
Rucha R. Jogchand et al. [12]	2017	Smart home automation	Control remote appliances in home automation	Dual tone multi frequency wireless module	• DTMF encoder • DTMF decoder • DTMF keypad	Control home appliances using mobile

TABLE 5.5 *(Continued)*

Authors	Publication year	Application type	Objective	Technology used	Methodology	Functional outcomes
Hinal Shah et al. [13]	2017	Smart home automation	Remotely monitoring and control home automation	Zigbee	• RF Transceiver controller • PIC24F microchip microcontroller • MPLAB IDE	Controlling and monitoring smart home using zigbee
Elsa et al. [14]	2017	Smart home automation	Design a security system for smart home automation	GSM	• R3 Board Omega 328 microcontroller • GSM Shield • Ethernet Shield	Design smart home security system with GSM mobile communication
Andria Procopiou et al. [15]	2019	Smart home automation	Identify DDOS attach in IOT network	802.11 Forcasting & Chaos theory	• forChaos algorithm for IOT device • DDOS attack simulation	Fast and accurate detecting DDOS attack in smart home iot network
Harsh Kumar et al. [16]	2019	Smart home automation	Remotely control electrical appliances	WI-FI	• Wifi-based microcontroller • Node MCU • Relays	Efficiently control electrical switches in smart home using remote system
Robert A. Sowah et al. [17]	2020	Smart home automation	Develop secure wireless home automation	OpenHAB REST API Wireless sensor network	• JSON web tokens • OpenHAB communication • AES • Android application	Provide secure control home appliances and wireless controlled system in smart home

TABLE 5.5 *(Continued)*

Authors	Publication year	Application type	Objective	Technology used	Methodology	Functional outcomes
Tran Anh Khoa et al. [18]	2020	Smart home automation	Develop smart light system for home automation	WI-FI WLAN	• Secure hash algorithm for enhancing security • ESP8266-12F micro controller deployed • Tracking and monitoring using webserver	Realtime remote monitoring system for smart lighting and enhancing security in smart home
Rizwan Majeed et al. [19]	2020	Smart home automation	Design a secure and intelligent home automation system	Block chain technology	• Machine learning algorithm support vector machine to taken intelligent decision • Authenticate and identify IOT device using block chain • Android application for smart home interface • SVM based classification	Design a secure and decision making ability to home electronic appliances
Nermeen A. Eltresy et al. [20]	2020	Smart home automation	Design an indoor monitoring system in home automation	Zigbee	• Design RF energy harvesting system • Power management system • A novel hybrid system for indoor environment prediction • IOT system interface design	Energy efficient Electromagnetic waves based indoor environment prediction

TABLE 5.5 *(Continued)*

Authors	Publication year	Application type	Objective	Technology used	Methodology	Functional outcomes
Anand Nayyar et al. [21]	2016	Smart agriculture	Design a smart stick assistant to monitoring the environment for agriculture	Wi-Fi Cloud computing	• ESP8266 WIFI module • Sensors, solar plate • Fetching live data using smart stick	Develop a stick for moisture temperature prediction successfully
Pratibha et al. [22]	2017	Smart Agriculture	Design a system to monitoring the temperature and humidity in smart farming	WiFi	• CC3200 single chip microcontroller • Send mms to given appropriate information to farmer	Develop a smart automation agriculture system
Karim Foughali et al. [23]	2018	Smart Agriculture	Design a system to prevent the disease in agriculture	WSN Cloud Computing	• Deploy cloud IOT & Wire-less sensor network • SIMCAST model to predict humidity • Light blight DSS	Identify the late blight disease in potatoes using cloud IOT
L.yang et al. [24]	2018	Smart agriculture	To detect chemical fertilizers & pesticides	LORA CLOUD	• Deploy Deep belief network • Apply SOFTMAX classifier • Wireless communication through LORA • Measure PH value, moistures etc using sensor nodes	Control and monitor chemical fertilizers in smart farming

TABLE 5.5 *(Continued)*

Authors	Publication year	Application type	Objective	Technology used	Methodology	Functional outcomes
Muhammad Ayaz et al. [25]	2019	Smart agriculture	Role of different IOT technologies in smart farming	WSN WIFI Cloud Bluetooth Zigbee	• Paper based survey analysis • Analyze different agriculture areas IOT technologies	Paper based analysis for prediction of crop maximization
Abhijit Pathak et al. [26]	2019	Smart agriculture	Design a system for water allocation for smart farming	Cloud computing	• Deploy cuckoo search algorithm • Sensors for prediction the PH value , moisture of soil & temperature	Prevent the water wastage in smart farming
Dinesh Manikandan et al. [27]	2019	Smart agriculture	Design a weather aware IOT system for smart agriculture	WI-FI WSN Cloud	• Deploy a agro gain platform • Path planning algorithm for UAV	Agro gain implements for precision agriculture
Abdellah Chehri et al [28]	2020	Smart agriculture	To reduce the network connectivity of smart digital farming	WSN	• Deploy underground wire-less channel • Randomly deploy sensor nodes • Install IOT embedded devices	Random or deterministic deployment of wsn to reduce network problem

TABLE 5.5 *(Continued)*

Authors	Publication year	Application type	Objective	Technology used	Methodology	Functional outcomes
Sohail Jabbar et al. [29]	2017	Smart healthcare	To provide interoperability between different heterogeneous devices for healthcare	Cloud computing	• Deploy semantic interoperability model • Semantic annotation • SWE framework • SPARQL queries	Tracking and monitoring human diseases through IOT-SIM
Vankamamidi S. Naresh et al. [30]	2019	Smart healthcare	Design lightweight healthcare monitoring system	RFID WSN	• Deploy IOT based RFID • HEC based algorithm • WBAN sensors	Store patients information and monitoring in real-time
Hui Li et al. [31]	2019	Smart healthcare	Design a IOT fog cloud framework	Cloud computing	• Deploy geo distributed fog layers • Design an encryption scheme for searching	Implement fog cloud based system for healthcare
Arshad Ahmad et al. [32]	2020	Smart healthcare	Design an energy efficient protocol for IOT healthcare	WSN LCX-MAC	• Increase throughput • reduction in delay & energy consumption through LCX-MAC • Extend X-MAC/BEB • Implement Markov model • Single hope network topology	Provide better efficiency in IOT healthcare network through LCX-MAC

TABLE 5.5 *(Continued)*

Authors	Publication year	Application type	Objective	Technology used	Methodology	Functional outcomes
Khizra Saleem et al. [33]	2020	Smart healthcare	Design a monitoring system for patients sleep quality	WLAN Cloud computing	• Deploy COS Sensors for monitoring patient's sleep • Forecast model & decision rules • Random forest algorithm for data classification	Monitor and analyze sleep pattern of patients
Rozita Jamili Oskouei et al. [34]	2020	Smart healthcare	Design a support system for Alzheimer's patients	WI-FI	• MQTT and WebSocket for information gathering • Nginx Web server • Implement REST API method algorithm for communication	Successfully develop a system for Alzheimer patients to track their health status
Kashif Hameed et al. [35]	2020	Smart healthcare	Design an intelligent healthcare system	WIFI Bluetooth	• Develop fuzzy neuron model (ARIC) • Deploy emergency sensors	Design medical DSS for patients data
Chong feng et al. [36]	2020	Smart healthcare	Design an energy efficient fog based healthcare system	WSN Cloud computing	• Implement fog nodes • Deploy sensors for patients data • Data analytics	Energy efficient healthcare sensor network

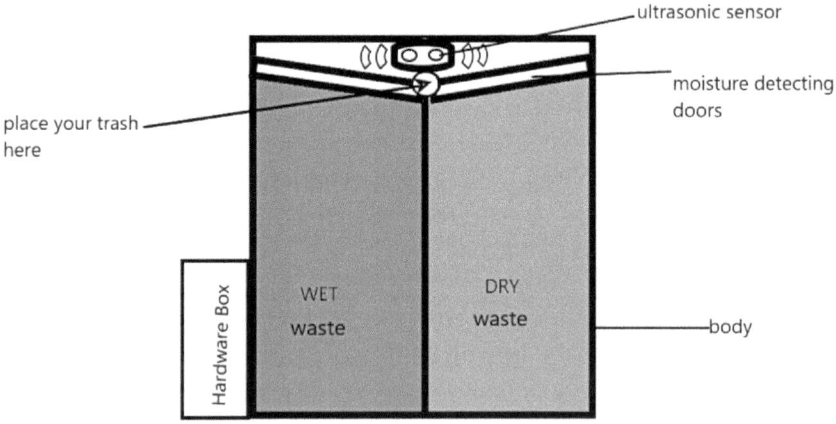

FIGURE 5.3 Smart dustbin functional units.

accordingly drop the waste into the intended partition. Once the trash is placed the sensor will detect the type of waste on the doors.

Components to need for this application:

1. Micro-controller board like Arduino Uno;
2. Ultrasonic sensor; and
3. Servo motor.

This smart dustbin application makes use of an Arduino board, moisture sensor, and IoT device, which assures better management of waste, and also helps in maintaining cleanliness everywhere in the city streets, roads, colonies, etc.

5.6.2 SMART HEALTHCARE

Artificial intelligence (AI) is a new upcoming emerging technology that has a vast impact on all human beings. Today all humans need a healthier life and for those, there is a proper understanding of their daily diet routine. The proposed smart application generally focuses on women's related problems which they do not easily share with anyone even if they have in pain and not also to doctors generally feel like embracement. This application act as a virtual assistant which predicts health issues from time to time and serves a better health result in front of the patient. This uses natural language processing for parsing and understanding what the human is saying. In a survey of healthcare and medical science, there is very shocking news that in India there is only one doctor over 2,500 patients and doctors work under heavy pressure, and many other critical issues are exposed. Thus, the proposed smart healthcare application is a chatbot that creates an electronic health record (EHR) report in which doctors easily predict and suggest medicine as per patient requirements. This virtual assistant is both conversational and speech based which is also easily used by a blind person. AI helps to predict an optimal solution to any problem in vast cases. In difficult and very critical operations, new disease observation may easily recognize and treat the patient with this chatbot application. Chatbot application can be integrated with google assistant to build it much better and easier to use, it reduces the cost of the project and does not create an unnecessary burden on the server. This smart chatbot application work to book an appointment, check health status, monitor health issues, remind for medicine, and other. It is used via mobile application or any high-profile website and used to learn with its past experience and training and add the command in its knowledge base and make it more valuable. The AI can predict the disease based on the symptoms and give a list of available treatments.

5.7 CONCLUSION

The internet of things (IoT) and its smart real-time applications have increased in popularity and production in the last few years. According to a project report on IoT, the economic value is rapidly increasing day by day and it goes approximately from the current $3.9 Trillion to $11.1 Trillion per year by 2025 [39]. This impact shows that there are approximately 50 billion devices are connected with the help of the network (internet). In the IoT, the IoT network is divided into two types of system groups. The first group, which has the objects (things) properties to provide high user familiarity, is called the things-centric system, and the second group, which has the functionality of IoT services and data processing tools, is known as a cloud-centric system [40]. This type of system (cloud-centric system) is mostly used because of the virtualization facility. It seems that the IoT is involved day by day in our life. This chapter first discusses the

IoT and its enabling technologies. We analyze a lot of literature surveys that use IoT-based smart applications like smart home automation, smart farming (agriculture), smart healthcare, and smart cities. Literature shows the different enabling technologies and different real-time applications which are done in previous years. In the last section, we summarize some challenges which occur in IoT networks.

KEYWORDS

- **electronic health record**
- **internet of things**
- **near field communication**
- **network**
- **personal area network**
- **sensors**
- **wireless**

REFERENCES

1. Linbo, Z., & Hua, W., (2017). Crowdsensing task assignment based on particle swarm optimization in cognitive radio networks. *Wireless Communications and Mobile Computing, 2017,* 9. Article ID 4687974.
2. Beltran, V., Skarmeta, A. F., & Ruiz, P. M., (2017). An ARM-compliant architecture for user privacy in smart cities: SMARTIE—Quality by design in the IoT. *Wireless Communications and Mobile Computing, 2017*, 13. Article ID 3859836.
3. Sajida, I., & Young-Bae, K., (2018). A novel indoor positioning system using kernel local discriminant analysis in internet-of-things. *Wireless Communications and Mobile Computing, 2018*, 9. Article ID 2976751.
4. Muhammad, A., Davide, M., Gabriele, P., Marco, T., Miguel, G., José, F., Joaquim, F., & Giuseppe, R. L., (2018). Real-time smart parking systems integration in distributed ITS for smart cities. *Journal of Advanced Transportation, 2018*, 13. Article ID 1485652.
5. Abd-Elhamid, T. M., (2018). An IoT architecture for assessing road safety in smart cities. *Wireless Communications and Mobile Computing, 2018*, 11. Article ID 8214989.

6. Francesc, A., & Ma Alsina-Pagès, R., (2019). Review of wireless acoustic sensor networks for environmental noise monitoring in smart cities. *Journal of Sensors, 2019*, 13. Article ID 7634860.
7. Daniel, M., & Benedict, O., (2019). Practical aspects for the integration of 5G networks and IoT applications in smart cities environments. *Wireless Communications and Mobile Computing, 2019*, 30. Article ID 5710834.
8. Yangchun, C., Guangyu, Z., & Chunyao, O., (2020). Application of financial cloud in the sustainable development of smart cities. *Complexity, 2020*, 11. Article ID 8882253.
9. Aggeliki, V., Georgios, S., George, C., George, S., & Phivos, M., (2016). User-adaptive and context-aware smart home using pervasive and semantic technologies. *Journal of Electrical and Computer Engineering, 2016*, 20. Article ID 4789803.
10. Noel, N. W. T., János, B., & Naoyuki, K., (2016). Weighted constraint satisfaction for smart home automation and optimization. *Advances in Artificial Intelligence, 2016*, 15. Article ID 2959508.
11. Aliyu, S., Yusuf, A., Abdullahi, U., Hafiz, B. M., & Lukman, A., (2017). Development of a low-cost GSM-Bluetooth home automation system. *International Journal of Intelligent Systems and Applications, 8*, 41–50.
12. Jogdand, R. R., & Bhupesh, N. C., (2017). DTMF based home automation system. In: *2017 International Journal of Engineering Science and Computing*.
13. Hinal, S., Vineeta, C., & Rashmi, S., (2017). *International Journal of Emerging Trends & Technology in Computer Science (IJETTCS), 6*(3).
14. Isa, E., & Sklavos, N., (2017). Smart home automation: GSM security system design & implementation. *Journal of Engineering Science and Technology Review, 10*(3), 170–174.
15. Andria, P., Nikos, K., & Christos, D., (2019). ForChaos: Real-time application DDoS detection using forecasting and chaos theory in smart home IoT network. *Wireless Communications and Mobile Computing, 2019*, 14. Article ID 8469410.
16. Singh, H. K., Verma, S., Pal, S., & Pandey, K., (2019). A step towards home automation using IoT. In: *2019 Twelfth International Conference on Contemporary Computing (IC3)*, Noida, India, pp. 1-5, doi: 10.1109/IC3.2019.8844945.
17. Robert, A. S., Dale, E. B., Dalton, C. O., Rexford, A., Godfrey, A. M., Owusu-Banahene, W., Gifty, B., & Sarkodie-Mensah, B., (2020). Design of a secure wireless home automation system with an open home automation bus (OpenHAB 2) framework. *Journal of Sensors, 2020*, 22. Article ID 8868602.
18. Tran, A. K., Le Mai, B. N., Hoang, H. S., Nguyen, M. T., Cao, H. P., Nguyen, T. H. P., Nguyen, V. D., et al., (2020). Designing efficient smart home management with IoT smart lighting: A case study. *Wireless Communications and Mobile Computing, 2020*, 18. Article ID 8896637.
19. Rizwan, M., Nurul, A. A., Imran, A., Yousaf, B. Z., Muhammad, F. M., & Muhammad, U., (2020). An intelligent, secure, and smart home automation system. *Scientific Programming, 2020*, 14. Article ID 4579291.
20. Eltresy, N. A., Dardeer, O. M., Al-Habal, A., Elhariri, E., Abotaleb, A. M., Elsheakh, D. N., Khattab, A., et al., (2020). *Smart Home IoT System by Using RF Energy Harvesting*. Preprints, 2020070280.

21. Nayyar, A., & Puri, V., (2016). *Smart Farming: IoT Based Smart Sensors Agriculture Stick for Live Temperature and Moisture Monitoring Using Arduino, Cloud Computing & Solar Technology*, 673–680. 10.1201/9781315364094-121.
22. Prathibha, S. R., Anupama, H., & Jyoti, M. P., (2017). IoT-based monitoring system in smart agriculture. *International Conference on Recent Advances in Electronics and Communication Technology.*
23. Karim, F., Fathallah, K., & Frihida, A., (2018). Using cloud IoT for disease prevention in precision agriculture. *Procedia Computer Science, 130.* 575–582. 10.1016/j.procs.2018.04.106.
24. Ling, Y., Sarath, B. V., Juan, Z., Xu, C. C., Ting, W., & Li, L., (2018). The development of an intelligent monitoring system for agricultural inputs based on DBN-SOFTMAX. *Journal of Sensors, 2018*, 11. Article ID 6025381.
25. Ayaz, M., Uddin, A., Sharif, Z., Mansour, A., & El-Hadi, A., (2019). Internet-of-things (IoT)-based smart agriculture: Toward making the fields talk. *IEEE Access*, 129551–129583. 10.1109/ACCESS.2019.2932609.
26. Hussain, S., Nahar, N., Pathan, R., Pathak, A., Tutul, A., & Abedin, Md. J., (2019). IoT-based smart system to support agricultural parameters: A case study. *Procedia Computer Science, 155*, 648–653.
27. Manikandan, D., Skl, A., & Sethukarasi, T., (2020). Agro-gain - an absolute agriculture by sensing and data-driven through IoT platform. *Procedia Computer Science, 172*, 534–539. 10.1016/j.procs.2020.05.065.
28. Chehri, A., Chaibi, H., Rachid, S., Hakem, N., & Wahbi, M., (2020). A framework of optimizing the deployment of IoT for precision agriculture industry. *ScienceDirect a Procedia Computer Science, 176*(2020), 2414–2422.
29. Sohail, J., Farhan, U., Shehzad, K., Murad, K., & Kijun, H., (2017). Semantic interoperability in heterogeneous IoT infrastructure for healthcare. *Wireless Communications and Mobile Computing, 2017*, 10. Article ID 9731806.
30. Vankamamidi, S. N., Sivaranjani, R., & Nistala, V. E. S. M., (2020). Secure lightweight IoT-integrated RFID mobile healthcare system. *Wireless Communications and Mobile Computing, 2020*, 13. Article ID 1468281.
31. Hui, L., & Tao, J., (2019). A lightweight fine-grained searchable encryption scheme in fog-based healthcare IoT networks. *Wireless Communications and Mobile Computing, 2019*, 15. Article ID 1019767.
32. Arshad, A., Ayaz, U., Chong, F., Muzammil, K., Shahzad, A., Muhammad, A., Shah, N., & Habib, U. K., (2020). Towards an improved energy efficiency and end-to-end secure protocol for IoT healthcare applications. *Security and Communication Networks, 2020*, 10. Article ID 8867792.
33. Khizra, S., Imran, S. B., Nadeem, S., Waheed, A., & Amna, A., (2020). IoT healthcare: Design of smart and cost-effective sleep quality monitoring system. *Journal of Sensors, 2020,* 17. Article ID 8882378.
34. Rozita, J. O., Zahra, M., Zohreh, B., & Khuda, B. J., (2020). IoT-based healthcare support system for Alzheimer's patients. *Wireless Communications and Mobile Computing, 2020,* 15. Article ID 8822598.
35. Kashif, H., Imran, S. B., Shabana, R., Waheed, A., & Akmal, K., (2020). An intelligent IoT-based healthcare system using fuzzy neural networks. *Scientific Programming, 2020*, 15. Article ID 8836927.

36. Chong, F., Muhammad, A., Arshad, A., Ayaz, U., & Habib, U. K., (2020). Towards an energy-efficient framework for IoT big data healthcare solutions. *Scientific Programming, 2020*, 9. Article ID 7063681.
37. Ali, S., Syed, K., & Mahmood, W., (2020). Smart home automation using IoT and its low-cost implementation. *I. J. Engineering and Manufacturing MECS, 5*, 28–36.
38. Singh, H. K., Verma, S., Pal, S., & Pandey, K., (2019). A step towards home automation using IoT. In: *2019 Twelfth International Conference on Contemporary Computing* (IC3) (pp. 1–5). Noida, India. doi: 10.1109/IC3.2019.8844945.
39. Manyika, J., (2015). *The Internet of Things: Mapping the Value Beyond the Hype*. McKinsey Global Institute.
40. Gubbi, J., Buyya, R., Marusic, S., & Palaniswami, M., (2013). Internet of things (IoT): A vision, architectural elements, and future directions. *Futur. Gener. Comput. Syst., 29*(7), 1645–1660.

CHAPTER 6

Intelligent University Monitoring System (i-UMS)

AKHILESH A. WAOO and ASHWINI A. WAOO

[1]*Department of CS/IT, FE&T, AKS University, Satna, Madhya Pradesh, India*

[4]*Department of Biotechnology, FLST, AKS University, Satna, Madhya Pradesh, India*

ABSTRACT

Intelligent University Management System (i-UMS), is a specially designed software application for precisely regulating a university/educational Institution's functions. It has real-time information available for the smooth functioning of each department of the university. Indian universities and academic institutions are facing many challenges during the pandemic of providing high-quality online education and management and monitoring of all aspects of a university. Students want user-friendly access to learning tools as well as admission, enrolment, and result-oriented information at their fingertips. i-UMS fulfills all these requirements with a great experience. Faculty members require online academic support for enhanced blended classroom teaching, and parents like to watch the progress of their children to secure their future. This can also be made possible by the i-UMS model. Overall, i-UMS is academically, technically, and socially helping the university to maintain the decorum and faith in the institution. This chapter focused on the models, features, significance, and applications of i-UMS.

Intelligent Sensor Node-Based Systems: Applications in Engineering and Science, Anamika Ahirwar, Piyush Kumar Shukla, Prashant Kumar Shukla, and Ruby Bhatt (Eds.)

6.1 INTRODUCTION TO I-UMS

In recent duration, universities play an important role in the education system. The main objective of the university is to monitor the efficiency of the educational system. The modified i-UMS is the expectation of the future COVID-19 world to upgrade this sector. The governments are approaching to launch new education policies towards a research-cantered approach along with the use of new technology in an efficient manner.

This chapter is a collection of practical modeling of academic structures in universities. Many practical problems are encountered in the student as well as the teaching and management level. This chapter will focus on two basic aspects while discussing the university intelligence system: (i) improvement of technology rather than physical infrastructure; and (ii) intelligent systems must not change the basic objective of education. i-UMS is a collection of various modules having various key features, as shown in Figure 6.1.

Key Features of i-UMS

- Easier Maintenance.
- Known by the Group Members.
- User (Student/Teachers/Management) Friendly.
- Unaffected to Change.
- Flexible in terms of Implementations.
- Flexible in terms of Data Searching & Report Generation
- Ease of Understanding.
- Ease of Modification and Extensibility.
- Secure and Connection Friendly.

FIGURE 6.1 Key features of i-UMS.

6.2 TRADITIONAL APPROACH

University has to follow a layered model for teaching-learning and evaluation of students. Online education will not be the only solution, but also it requires modification in the teaching process. Well-known technologies and hardware like intelligent surveillance systems (ISS) [1], intelligent gadgets, virtual lab, intelligent evaluation, motion sensor surveillance system, dynamic allocation of resources, ITS (intelligent transportation system) [3], and artificial intelligence (AI) makes it possible. Even, the ISS will provide integrated security of information in heterogeneous environments [2].

The present traditional education system consumes time in manual activities, lengthy paperwork, and is very rigid to follow static structure. Therefore, it needs to develop an i-UMS to overcome these problems. The current education system is not designed according to the creativity of the student/candidate/facilitator. The system suppresses the practical or laboratory approaches and gives priority to the theoretical concepts.

6.3 I-UMS – REQUIREMENT AND ANALYSIS

Requirement and analysis are preliminary stage tasks before software development of i-UMS which links between system engineering and vigorous requirements from the users. Analysis of essential parameters to collect overall idea about the system and also the drawbacks of the existing scenario. This analysis determines the factors that affect the structural organization of the entire i-UMS.

Among various platforms, the Ubuntu Linux Operating system is the best choice for University management because of viruses' issues and its freeware. i-UMS can also run under another operating system like Windows, Mac, etc. The initial stage of any software development is to analyze what problems to be solved? And what are the measures under consideration in the existing traditional scenario? The analysis process finds out user and system requirements along with the cost of software development, implementation, and maintenance (as shown in Figure 6.2).

Factors Affecting Development Cost of i-UMS

- The complexity and features of the i-UMS
- The platform on which it will run
- Time of development of i-UMS
- Security issues regarding i-UMS

FIGURE 6.2 Factors affecting development cost of i-UMS.

There is no average price for building a University Monitoring System considering the numerous factors. According to requirements, it can be variable. It is essential to clearly and define the specific objective, before acquiring the University Monitoring System. For that critical analysis of various factors such as academic processes including admission and enrollment, class attendance, assignment submission, discipline, fee payment, scholarship, and mark sheet as well as degree distribution should be conducted. There is a need to define specific improvements in any academic process for the solution to any existing problem.

6.4 I-UMS MODEL

In a perfect world, the i-UMS model should be completely automated. This chapter presented a viewpoint of the design and development of integrated intelligent information and control systems to fulfill functional requirements. Intelligent University management is responsible for the production of specialized software, such as academic software, which is a combination of tools that are responsible for automation and monitoring of the academic journey of the students. It is a need of time to use digital technologies in education, like industries to address modern trends. University management software is a smart way to improve productivity among management, teachers, and students. University management software development will not only boost productivity but also decrease overhead costs and simplify the educational procedure. This chapter will ensure the use of intelligent UMS with its advantages and categories.

i-UMS can combine various University operations essential for the educational process with automation. This functionality can be divided into different sections depending on the workflow of the university, as shown in Figure 6.3. i-UMS model has the following basic modules:

- Student information module;
- Department module;
- Examination module;
- Student placement and HR module;
- Finance and accounting module;
- Inventory and library module.

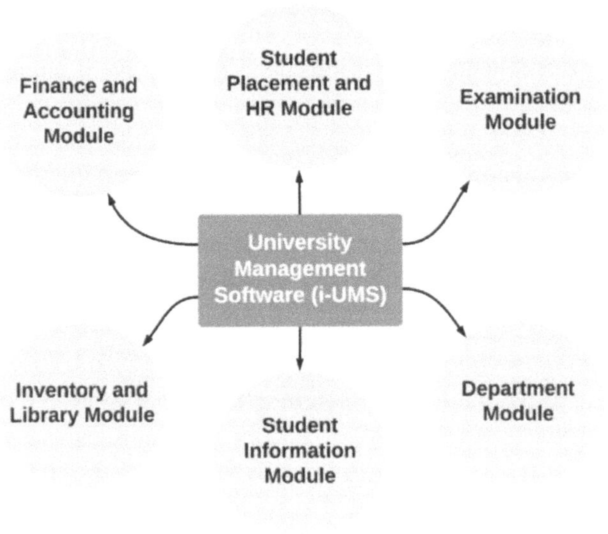

FIGURE 6.3 Modules of i-UMS.

6.4.1 STUDENT INFORMATION MODULE

The very important module of the University Monitoring System is the student information system which is a collection of student details. The student information system has directly interacted with around all the modules of i-UMS, as the student is the main entity in the education system. The student module has the following sub-modules apart from the main module interaction, such as the alumni module, online project, assignment submission module, and many more:

1. **Alumni Module:** Alumni association plays a significant role in the future of any university. It helps to organize alumni registration and discussion forum for all alumni [4]. It also includes a job referral system where any member can refer jobs for new students at the university and actively engage all the members to participate in the event and job communication purposes.
2. **Online Project/Assignment Submission:** These modules are responsible for collecting all types of assignments for remote students and monitor their status of work from anywhere they present. Students can submit their all types of assignments/projects via the online submission mechanism.

6.4.2 DEPARTMENT MODULE

University is a collection of various departments, and it requires a centralized interconnected mechanism to manage faculty information and course information along with the regular class schedule, student attendance, timetable management, subject management for smooth University functioning. The department module keeps track of student's information and their progress, courses, and admission details, teaching staff, and subject allotment. An Intelligent University Monitoring System is used for taking attendance via smartphone to automatically detect class attendance during online classes.

1. **Event Management System:** i-UMS can be used for registration purposes and all data of upcoming events can be uploaded on websites as well as previous event recordings can also be made available for everyone. It can result in better management and organization in keeping records over the years.
2. **Progress Monitoring:** The progress of each student can continuously monitor the status of the assignment and all tests. It will give the accurate grading of students according to their performance.
3. **Online Admission:** Students can collect information about all these courses available in the university, job opportunities, and the scope of a particular subject through an online admission system and they can also confirm their admission through an online admission software system and join online practices.
4. **Teaching and Learning:** The teaching and learning module is responsible for credit-based system execution. This includes the teaching pedagogy along with the various learning activities with the evaluation of students along with course-outcome. It will be kept as a record for accreditation purposes.

6.4.3 EXAMINATION MODULE

Examination software monitor and manage the process of conducting examination and result generation along with keeping records of student and the status of backlog papers and examination fee details. Examination management help to carry out exams on smartphones, computers, and laptops. It will also provide a medium for evaluating the answer sheet of the students. It is necessary to evaluate the subjects and generation of automatic results along with degree distribution. It can keep track of all issued mark sheets and degrees according to the year.

6.4.4 STUDENT PLACEMENT AND HR MODULE

Job placement for students is a very crucial aspect of any successful academic institution. In the traditional system, manual data collection is carried out by students, filling the form and then the compilation of data in an excel sheet. An intelligent placement system can manage the complete placement module of the university via the software application. The administrator can verify the authenticity of the student's data along with previous academic records and current graduation details. These records can be used to forward for any campus drive.

The human resources (HRs) module deals with staff recruitment and maintains staff records. It keeps and simplifies accession of employee details along with job hours, salary structure, faculty designations, role, and responsibilities.

6.4.5 FINANCE AND ACCOUNTING MODULE

The finance or accounting department is the pillar of every organization. It is mandatory to do the transactions, in the fastest and secure manner. Intelligent UMS can automate the budget along with expenses payment details and helps to generate all types of statistics related to finance.

6.4.6 INVENTORY AND LIBRARY MODULE

Smart i-UMS helps to keep track of stock storage materials, used, and unused items, product transition with various departments. The library module can keep track of book records for the library. Smart UMS can use RFID mechanisms to track the book and its location automatically.

6.5 CONTENT DEVELOPMENT

Online learning is only web-based instructions (WBI) for synchronous and asynchronous exchanges of the resource. It is a type of delivery method used in distance education [5]. Content development is a very significant task in the online learning process. e-learning content development is possible through the use of various online content development tools:

1. **Text Editors:** The textual content of the syllabus can be developed through MS word or PDF files. It can be used for giving instructions, assignments, and can be displayed on a web page or as a blog in the learning management system.
2. **Graphics and Images Tools:** Some important mechanisms of the syllabus can be developed through graphs and images using photoshop/CorelDRAW/paint in different formats like JPG or PNG in a reduced size for web uploading. During this copyright, conditions must be followed.
3. **Presentation Tools:** Slide presentations can be used in online classes it is the most powerful and efficient way for preparing online lessons. Some of the best presentation tools include Microsoft PowerPoint, Visme, Prezi, Canva, and Google Slides.
4. **Screen Recorder and Screen Capture:** Some software helps to capture the screen on how to record the screen that can be used in an online class to demonstrate some practical skills. Some of the best screen recorders are FreeCam, Free Screen Video Recorder, ShareX, CamStudio, Ezvid, TinyTake, Filmora Scrn, etc.
5. **Audio-Video Tools:** Audio-visuals are extremely effective in demonstrating practical protocols in online teaching. Some of the audio editors are Audacity, SOUND FORGE, ocenaudio, Adobe Audition CC. Similarly, a list of video editors can include Adobe Premiere Pro, Final Cut Pro X, Lightworks VSDC Free Video Editor, Pinnacle Studio, iMovie, etc.

6.5.1 ONLINE CONTENT DESIGNING PROCESS

For an enjoyable experience of teaching for the instructor and an easy understanding of students, an online course must be well designed. For appropriate designing of content following points must be taken into consideration:

- Focus on learning objectives;
- Content must be consistent and in a sequenced manner;

- Comprehensive development of the syllabus;
- Learning objectives must be combined with activities and assessments;
- Conceptualize the content;
- Follow some copyright issues and accessibility standards.

In the era of digitization of the world internet and the World Wide Web help learners to access information easily and effectively. Traditional classrooms were transformed into global classrooms in the 21 century. The advancement in technology brings various innovative digital devices and online tools to have an efficient improvement in education. Digital content and devices improve enjoyable learning as well as educational opportunities. Online educators should follow standards of e-content by using advanced digital resources for teaching and learning as shown in Figure 6.4. Teachers must have proper pieces of training and skills in e-content development.

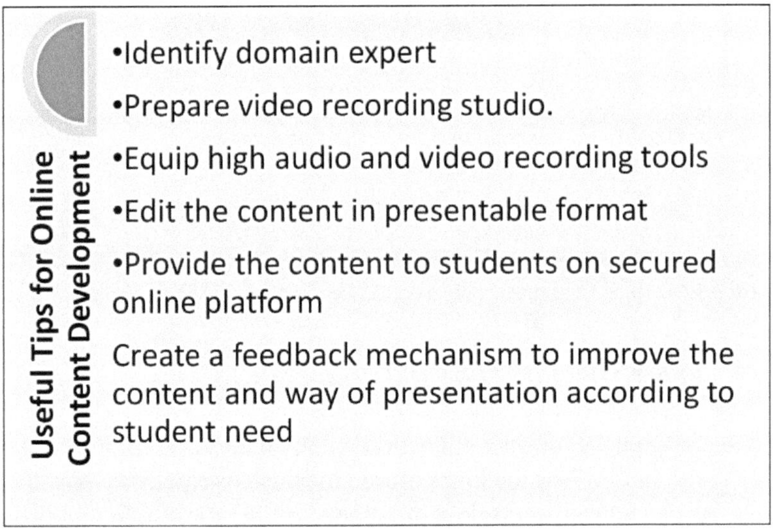

FIGURE 6.4 Useful tips for online content development.

Online content development required a much higher standard of systematic training for instructors and faculty members [6]. University management can concentrate on the professional development of all their professionals along with research and development activities. It is essential to always adopt all-new available technologies and must be ready to adopt new methods of teaching and learning. Online teaching and learning methods are not alternatives to regular education mechanisms but they work like a catalyst for change.

The objective of the university is to develop knowledge and skillsets for their student's using hybrid blended and online learning with the collaboration of a knowledge base society. Knowledge cloud of teacher or moderator builds with subject input experience in the concerned field and its application for generating, organizing, and convergence of new ideas (Figure 6.5).

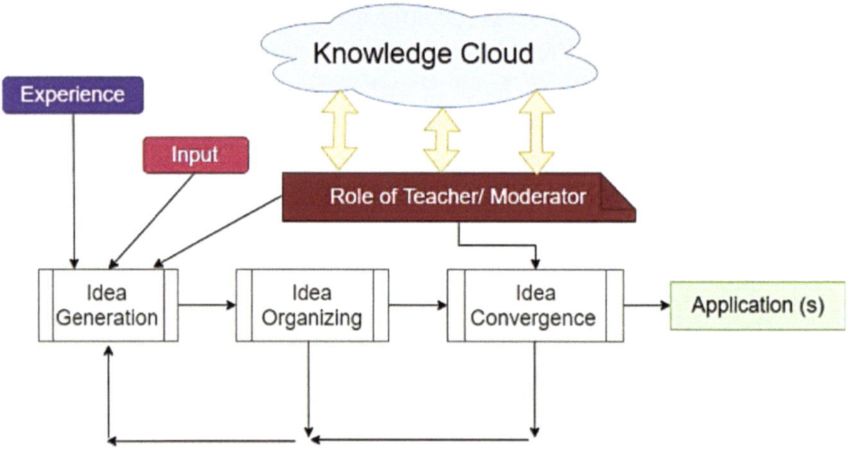

FIGURE 6.5 Role of teacher/moderator in online content designing process.

6.5.2 *EFFECTIVE CONTENT DELIVERY TOOLS*

The main goal of online education is to achieve learning objectives effectively and collect feedback from students. Interactivity plays a vital role in online education. Below mentioned online tools help to deliver the educational content effectively and collect the feedback of students:

1. **Edpuzzle:** An interactive tool for Crop, customize, and remix online video content.
2. **Screencast-O-Matic:** Creating and sharing high-quality screencasts without breaking the bank.

3. **TalkingPoints:** Multilingual texting tool for enhanced communication.
4. **Remind:** A highly effective messaging tool for keeping students and parents connected with online classes [7].
5. **Socrative:** This tool allows teachers to create online games and quizzes. Students can solve these quizzes by using mobile or laptop devices. Teachers can analyze the results of the activities done by the students.
6. **Projeqt:** To create dynamic flight and multimedia presentations with the embedded interactive map, online quizzes, and videos. It can share with the students during academic presentations and online class sessions.
7. **Microsoft Whiteboard:** This tool gives an online board for the teachers to draw some diagrams with explanations and text boxes. It is easy to embed any presentation slides or PDF content.

6.6 COLLABORATIVE LEARNING

The collaborative learning approach actively participates the learner in the process of learning. It helps to synthesize the information and concepts of learners while working in groups or with collaborative interactively to understand the concepts. Online collaborative learning integrates theories that focus on conversational learning [8], deep learning [9, 10], knowledge construction [11], and the development of academic knowledge [12]. Collaborative learning is a different approach for teaching and has a more objective-oriented teaching-learning process in online collaborative learning technology. It is used to increase and improve the communication between teachers and students for the development of an outcome-based learning process.

Online education environments give real challenges to students in the process of learning. These challenges will be overcome by adopting special technological tools and pedagogy in online teaching to continuously monitor the progress of the learner. The following aspects have to be taken into consideration while online teaching:

- Specific software or technology can be used for online teaching;
- Standard guidelines for students about online software;

- Student orientation for the subject knowledge and awareness about the scope of the subject;
- Outcomes of specific goals are specified and understood by the students;
- Choice Based system for learning topics;
- The clarity in learner roles and expectations;
- Continuous monitoring of individual learners and response mechanism.

6.7 CONTROL AND MONITORING

It is necessary to implement the i-UMS with administrative features for continuous surveillance of the online teaching-learning process. The thought is to empower the educator to see all the activities of students and teachers on screens as well as to check connectivity within the scheduled duration. This helps in observing students along with their performance. The system should detect the screen if there is no activity from the teacher and student in an online class in a certain configurable period [13]. Laboratory work can be done in three different manners – practical demonstrations, individual work (Hands-on Practice) by students on projects/ Lab tasks, and result testing. i-UMS should provide a solution as a fully automated integrated information and control system that builds on top of several practices, tools, and applications for monitoring and network access control. The teacher should monitor the lab content and defines the lab protocol according to the intelligent education system.

It will be a better option to save video recordings of lab practices in i-UMS. Laboratories are scheduled according to the timetable by using stored videos of lab practices that will help students to take references in the future, and also, they will not depend on the physical availability of labs.

6.8 SECURITY ISSUES IN I-UMS

In online learning, security means that learning resources are available to only authorized users according to their requirements. Access Control Policies must be used to ensure control of information systems and information processing facilities, and all access rights are properly authorized.

Proper security measures should be applied to remote accesses [14]. Confidential control access should be there in case of student personal as well as financial information are stored. Specific password controls should be enforced in the access protocol. Security issues in information collection and communication always deal with the six basic principles, namely Confidentiality, Integrity, Authenticity, Accountability, Non-Repudiation, and Availability, as shown in Figure 6.6. It is necessary to design i-UMS with all the above-mentioned principles of information security to make it more secure and reliable [15].

FIGURE 6.6 Principles of information security.

The University Monitoring System ensures that all of its staff and students should have an awareness of information security and also understand various policies about the security of any type of information in the academic record [16]. Some sort of training and preparing sessions must be there to convey information security policies to all the staff and students [17].

6.9 SUMMARY

Technologies like artificial-intelligence internet GPRS Wi-Fi, and intelligent use of sensors help to make super i-UMS. i-UMS ultimately serves as a smart tool for the administration of the university through major modifications such as paperless work, faster results, and efficient and accurate data management. i-UMS establishes operational efficiency and creates self-service in a user-friendly manner. Secure i-UMS will be the smart solution for the new education system.

KEYWORDS

- **blended learning**
- **education**
- **human resources**
- **intelligent surveillance system**
- **intelligent transportation system**
- **i-UMS**
- **software**
- **web-based instructions**

REFERENCES

1. Qian, H., Wu, X., & Xu, Y., (2011). *Intelligent Surveillance Systems, Control and Automation: Science and Engineering.* Springer Science & Business Media. ISBN: 9400711379, 9789400711372.
2. Castro, J., Delgado, M., Medina, J., & Ruiz-Lozano, M., (2011). Intelligent surveillance system with integration of heterogeneous information for intrusion detection. *Expert Systems with Applications*, *38*(9), 11182–11192. ISSN: 0957-4174, https://doi.org/10.1016/j.eswa.2011.02.165.
3. Kyriakides, E., & Polycarpou, M., (2014). *Intelligent Monitoring, Control, and Security of Critical Infrastructure Systems.* Springer, ISBN: 3662441608, 9783662441602.
4. Jain, S., Bhosle, V., & Sah, L., (2017). Smart university-student information management system. *International Conference on Smart Technology for Smart Nation* (pp. 1183–1188).
5. Khan, B., (1998). Web-based instruction (WBI): An introduction. *Educational Media International, 35*(2), 63–71, doi: 10.1080/0952398980350202.
6. Bates, A., & Sangrà, A., (2011). *Managing Technology in Higher Education: Strategies for Transforming Teaching and Learning.* San Francisco: Jossey-Bass/ John Wiley & Co.
7. Blog, https://www.commonsense.org/education/top-picks/best-tools-for-virtual-and-distance-learning (accessed on 21 December 2022).

8. Pask, G., (1975). *Conversation, Cognition and Learning.* Amsterdam/London: Elsevier.
9. Marton, F., & Saljö, R., (1997). Approaches to learning. In: Marton, F., Hounsell, D., & Entwistle, N., (eds.), *The Experience of Learning*. Edinburgh: Scottish Academic Press.
10. Entwistle, N., (2000). Promoting deep learning through teaching and assessment: Conceptual frameworks and educational contexts Leicester UK: TLRP Conference.
11. Scardamalia, M., & Bereiter, C., (2006). Knowledge building: Theory, pedagogy and technology. In Sawyer, K., (ed.), *Cambridge Handbook of the Learning Sciences*, New York: Cambridge Univ. Press.
12. Laurillard, D., (2001). *Rethinking University Teaching: A Conversational Framework for the Effective Use of Learning Technologies*. New York/London: Routledge.
13. Ajanovski, V., (2015). *Access Control and Monitoring for Campus Computer Labs*. Best practice document, produced by the MARNET-led working group on network security and monitoring, (FCSE, UKIM).
14. Simon, K., & Cheung, (2014). Information security management for higher education institutions. In: Pan, J. S., et al., (eds.), *Intelligent Data Analysis and Its Applications: Advances in Intelligent Systems and Computing* (Vols. 1, 297, No. 11, pp. 11–19). doi: 10.1007/978-3-319-07776-5_2.
15. Adams, A., & Blandford, A., (2003). Security and online learning: To protect or prohibit. *Usability Evaluation of Online Learning Programs*, 331–359.
16. Harasim, (2012). *Harasim's Pedagogy of Group Discussion*, 95.
17. Harasim, L., (2012). *Learning Theory and Online Technologies*. New York/London: Routledge.

CHAPTER 7

Intelligent Home

LALIT PUROHIT and MANOJ DHAWAN

[1]*Department of Information Technology, Shri Govindram Seksaria Institute of Technology and Science, Indore, Madhya Pradesh, India*

[2]*Department of Information Technology, Shri Vaishnav Vidyapeeth Vishwavidyalaya, Indore, Madhya Pradesh, India*

ABSTRACT

In the era of the internet of things (IoT) and smart things, a huge growth in the smart devices is observed for intelligent home (IH). The smart devices are considered as indispensable and inseparable part of IoT. The four-layer architecture of IoT includes smart devices, electronics devices as its core/innermost layer. Further, for realization of the IH concept, IoT provides all necessary infrastructures. Various services offered in a IH are implemented with the help of web service technology. With the increasing use of web services in IH context, multiple web services offering identical functionality are available. Selection of the most appropriate web service to obtain the desired functionality is a fundamental problem in IH perspective. In this work, a IH study is discussed to understand the role of web services in IH realization.

7.1 INTRODUCTION

The intelligent home (IH) is the today's need of the human. With the advancement in the field of internet of things (IoT), the concept of IH

Intelligent Sensor Node-Based Systems: Applications in Engineering and Science, Anamika Ahirwar, Piyush Kumar Shukla, Prashant Kumar Shukla, and Ruby Bhatt (Eds.)

is refined and improved according to the need of the people. Father of IoT was Kevin Ashton, who coined the term IoT the year 1999. IoT is the collection of many interconnected objects, devices, services, which can communicate through wired/wireless mode and share the data, and information to achieve particular goals or applications without any human intervention [1]. The IoT is the creation of network-connected devices to reach a new level of autonomy. IoT, or IoT, refers to the billions of physical devices around the world that are now connected to the internet and capable of collecting and sharing data, as shown in Figure 7.1. The IoT is making the things easier and more flexible [2].

The IoT is a rising technology across the globe, which helps to connect sensors, transportations, buildings, industries, infrastructures, hospitals, and users (consumers) through internet connectivity. This type of architectures leads to IH, smart cities and smart world.

There are many applications of IoT in almost all fields. A few applications are as follows, Smart parking systems, Electromagnetic level detection system, Structural Health monitoring system, Urban noise maps, Smartphone Detection, Traffic congestion, Smart lighting system, Waste Management system, and Smart roads.

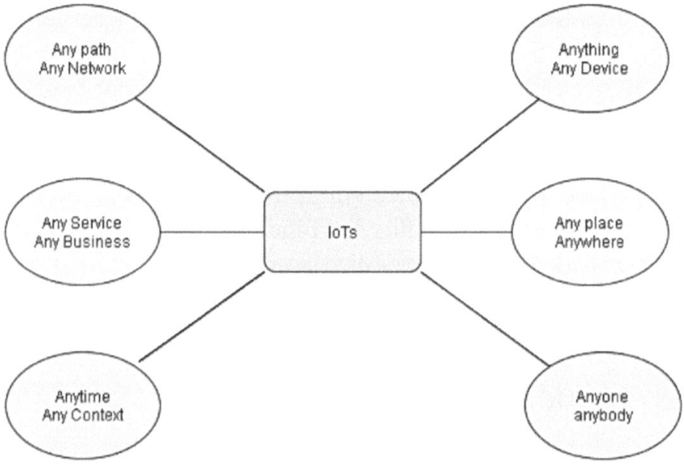

FIGURE 7.1 Internet of things.

The IoT includes a developing number of shrewd interconnected gadgets and sensors that are frequently non-meddling, straightforward, and imperceptible. The correspondence among these gadgets just as with related administrations is relied upon to happen whenever, anyplace, and it is much of the time done in a remote and autonomic way. Likewise, the administrations become substantially more decentralized and complex. In this manner, to manage the multifaceted nature, IoT design is required. Architecture in this setting is described as a structure for the detail of a framework's physical segments and their utilitarian affiliation and course of action, its operational principles and procedures, similar to data plans used in its movement [3]. The IoT architecture is presented in Figure 7.2 and is derived from Ref. [4].

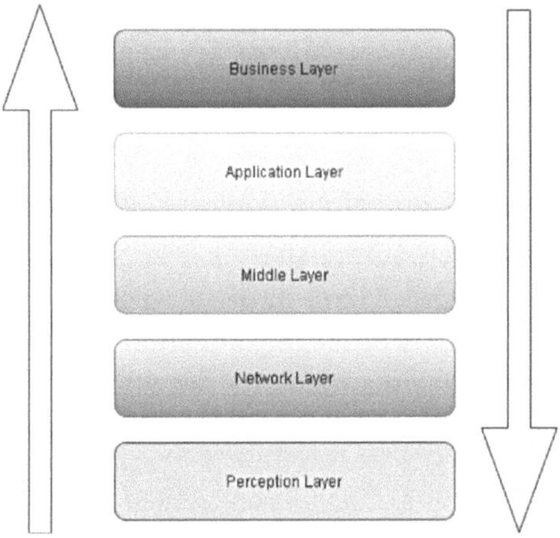

FIGURE 7.2 IoT layered architecture.

1. **Physical Layer:** The perception layer is the physical layer, which has sensors for sensing and assembling data about the environmental condition. It senses some physical movements or identifies other smart objects in the environment to communicate. The physical layer consists of constrained and unconstrained devices. The constrained devices are end nodes with sensors/actuators that can handle a specific application purpose and are less in power and less in memory. Unconstrained devices are those devices which have a high power and are huge in memory [5].
2. **Network Layer:** It is also known as communication layer. It acts like a bridge between the physical layer and the application layer.

It transfers the information collected by physical devices like sensors to the application layer. The medium for the transmission can be wireless or wired. The network layer uses transmission media, like 4G, 3G, 2G, fiber-optic, and short-range communications for correspondence the information over the public network. It is the responsibility of the communication layer for interfacing the smart things, network devices and networks to each other. The main function of this layer is data routing and transmission to different IoT hubs and devices over the internet. The cloud computing platform, Routing, Switching, Internet Gateways, and other devices act in this region using Bluetooth, ZigBee, Wi-Fi, LTE, 3G, etc. Data aggregation, Data filtering, and transmission also take place at this layer [6].

3. **Middleware Layer:** The devices in the IoT system may create different kinds of services when they are associated and communicated with others. The middleware layer has two basic works, including service administration and store the lower layer data into the database. More finished, this layer has the ability to recover, measure, figure data, and afterward consequently choose dependent on the computational results [4].
4. **Application Layer:** It is liable for comprehensive applications the board dependent on the handled data in the Middleware layer. The IoT applications can be smart postal, smart heath, smart vehicle, smart glasses, IH, smart autonomous living, smart transportation, etc. [4].
5. **Business Layer:** It works cover the entire IoT applications and administrations the executives. It can make reasonably graphs, plans of action, flow chart, administrative report, and so on dependent on the measure of exact information got from lower layer and powerful information investigation measure. In light of the great examination results, it will support the utilitarian directors or heads to make more precise decisions about the business methodologies and road maps [4].

Today, 55% of the total population lives in urban zones, an extent that is relied upon to increment to 68% by 2050. The slow move in home of the human populace from rural to urban territories, joined with the general development of the total populace could add another 2.5 billion individuals to urban regions by 2050 [7].

One other researcher claims that, due to inconsistent definitions, even these numbers underestimate the true extent of urbanization. As of 2018, an estimated 84% of the world's population already lives in urban areas, based on geospatial technology using high-resolution satellite images [8].

With over 55% of the world's population now in cities, significant strains are placed on city resources and infrastructures. The use of information and communications technologies (ICT) to modernize cities promises to create Smart Cities that mitigate the impacts of increased city populations while improving the quality of life (QoL) for all inhabitants [9].

The growth of the number of applied technology and variety of devices that are collecting data is implausibly rapid. A study by Cisco [10] expects that the number of Internet-connected devices will be 50 billion by 2020 (*refer* Figure 7.3).

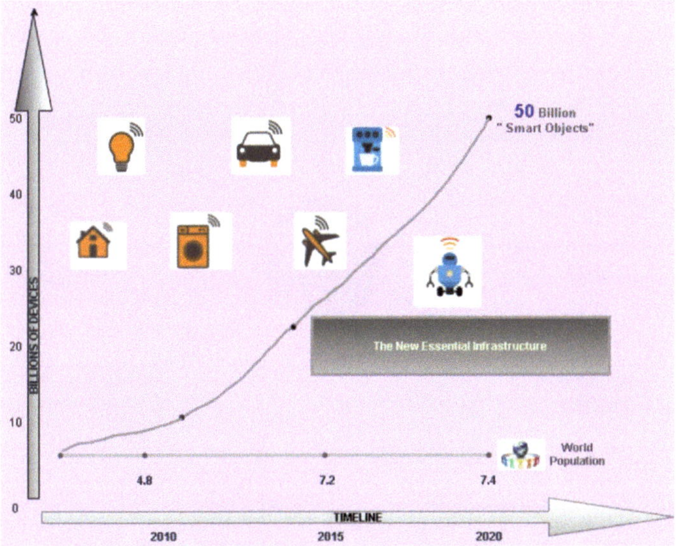

FIGURE 7.3 Internet-connected devices and future evolution.

The fundamental requesting is what is determined by a 'Shrewd City.' The most ideal reaction is, there is no typically seen significance of a shrewd city. It prompts different things to different people. The conceptualization of Smart City, therefore, contrasts from city to city and country to country, subordinate upon the degree of movement, essentialness to change and change, resources, and needs of the city occupants [11]. The conceptualization of Smart city shifts from one nation to another, contingent upon numerous viewpoints, including the degree of development of the nation. A smart city in America or Europe will have different characteristics compared to a smart city in India.

According to Monzon, "Smart city" concept coined towards the end of the 20 century. It is rooted in the implementation of user-friendly information and communication technologies developed by main industries for urban plots. Smart cities are forward-looking, progressive, and resource-efficient while presenting at the same time a high QoL; they enhance social and technological innovations and connect existing infrastructures [12]. In the beginning, the utmost goal of smart cities was to improve the QoL of urban citizens [13].

The overview of a typical smart city is illustrated in Figure 7.4.

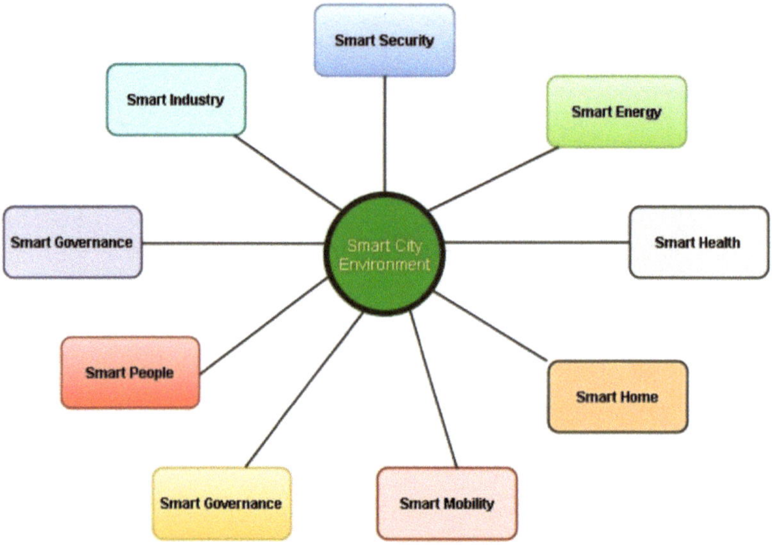

FIGURE 7.4 Overview of a basic smart city environment.

1. **The Smart Features of Smart Cities:** We can classify the smart features of smart cities on the basis of Mobility, Transportation, Services, and Infrastructure Connected.

2. **Smart Parking:** Real-time observing of parking spots accessible in the city making occupants ready to distinguish and the nearest accessible spaces.
3. **Traffic Solutions:** Decrease in gridlock and expanded income from dynamic estimating could be a portion of the advantages just as less complex obligation regarding traffic superintendents perceiving rebellious use.
4. **Intelligent Transportation System (ITS):** Smart streets and clever expressways with notice messages and preoccupations as indicated by atmosphere conditions and an unforeseen occasion like mishaps or gridlocks.
5. **Connected Learning:** Improvements in instructor usage decrease in instructional supplies, profitability improvement, and lower costs are instances of advantages that may pick up from letting electronic assets convey information driven, real and collective learning experience to bigger gatherings.
6. **Smart Lighting:** Intelligent and whether versatile lighting in streetlights.
7. **Waste Management:** Detection of garbage levels in holders to improve the trash collection courses. Trash bins and recycle bins with RFID labels permit the sanitation staff to see when trash has been put out. Possibly "pay as you through" program would assist with diminishing trash waste and increment reusing endeavors [15].
8. **Smart Irrigation of Public Spaces:** Maintenance of parks and yards by covering park water system observing sensors in the ground remotely associated with repeaters and with a remote door association with the web.
9. **Smart Water:** Leakage detection, Desalination, Water automation system, Quality Monitoring, and Supervisory Control.
10. **Smart Energy:** Concentrated solar, Solar PV, Smart grid, Smart Monitoring, District cooling, and Smart storage.
11. **Smart Buildings:** Building automation and management, building efficiency monitoring, smart storage.

12. **Smart Health:** Healthcare requires truly elevated measure upgrades; patient dies in the ambulance because of the absence of specific services, network, and accessibility. Profoundly prepared emergency vehicle services are required with the ongoing help of specialists 24×7. So, Smart Health is necessary for the Smart City.
13. **Intelligent Home (IH):** It can be defined as an intelligent, sustainable home which involves the use of information and communication technologies (ICTs) and other means to improve QoL, the efficiency of operations/services, and competitiveness [3]. Offered services in the context of IH include automation of home device operations as per the personalized need of individuals, changing room temperature, home security, Smart door with person identification system, etc.

7.2 THE ROLE OF WEB SERVICES IN IOT AND INTELLIGENT HOME (IH)

The heterogeneous technologies currently used in the IH realization also impose the problem of non-interoperability. The problem can be resolved by the use of IoT infrastructure that can act as a technical backbone for IH [18]. The IoT infrastructure acts as a source of information in IH environment. Devices generate data specific to the home environment. Data can be securely retrieved from the source, stored, and sorted in real-time to feed to various applications. These applications operate together to combine data and generate enriched information. The information can be used by tailored response services designed to improve the QoL of individual. Thus, IoT provides the necessary infrastructure to implement the IH concept.

In the IoT architecture (*refer* to Figure 7.2), web services exist at the application layer and are used for realizing applications for IH [19]. Web services act as binding between the IoT infrastructure and application to access the IH resources. The use of web services at the application layer is having advantage of being flexible, interoperable, and extendable to IoT devices, among others. The web service offering smart services to realize IH concept has attracted many service providers. As a result, a lot of functionally similar web services are available to perform the desired task. In order to ensure the efficiency of smart applications developed using web services, the task of web service selection must be efficient. QoS information can be used to compare, prioritize, and select web services. Moreover, when a number of functionally similar web services is significantly large in number, the task of selection of desired web service based on QoS also become difficult and time-consuming. This lead to defeat the purpose of IH to offer value-added services in real time [20]. Therefore, the web service classification can be employed for efficient selection of web service.

7.3 THE ROLE OF WEB SERVICES IN SMART CITIES

Web services act as a key technology for Internet of Things. Since the backbone of smart city is IoT, thus th Web Services play a very vital role for development of IH. With the increasing use of web service technology in software development and multifold advantages, quite a good number of service providers got attracted to offer their business functionality as a web service. IH is one such example. The IH web service allows users to intelligently manage their homes. The IH includes smart door alarms, IH applications, smart electrical appliances, etc. Figure 7.5 shows an IH application scenario using the IoT infrastructure of a smart city. The IH application case study is presented next.

A person completes all the assigned office work and is about to start from home from the office. Before starting, she used the IH application to communicate with the IH devices. The IH application interacts with the cab booking service, traffic management service, weather service, and smart traffic management web service to determine the best route to reach the home of the person. Based on the data from these web services, the approximate time for reaching the home of the person is determined. According to this calculated time, the smart AC and Coffee maker web services are coordinated to automatically manage the room temperature and prepare the coffee/tea, respectively. Also, the intelligent light web service is also given input to automatically get started upon arrival of the person as per the lighting condition. Clearly, from this scenario, web services play an important role in automatically processing the information. The desired goals of the person by effectively utilizing the resources. This also saves a lot of efforts and ensure the desired comfort of the end user. But many times several functionally equivalent web services are available to meet the functionality desired by the applications/end user. For example, in the IH application scenario, multiple service providers offer cab booking service, traffic management service, weather service, and smart traffic management web service. The multiple services offering the same functionality increases system availability and reliability however, the complexity of the system to select one web service get increase. Further, among the functionally similar web services, all the web services may not satisfy non-functional (QoS) requirements. Thus, to simplify the selection process of web services, QoS parameters are useful and can be used to group the web services with similar QoS requirements. For this grouping task, the classification techniques are useful. In this chapter, an empirical study of six classification models is considered. The classification models included in the study are neural network (NN), C4.5, decision table (DT), and random decision forest (RDF).

7.4 EMPIRICAL STUDY OF SIX CLASSIFICATION MODELS USED FOR WEB SERVICES CLASSIFICATION

For classifying web services, various classification models are available. Two best learning models, each from – decision tree-based, rule-based, and function-based categories are chosen.

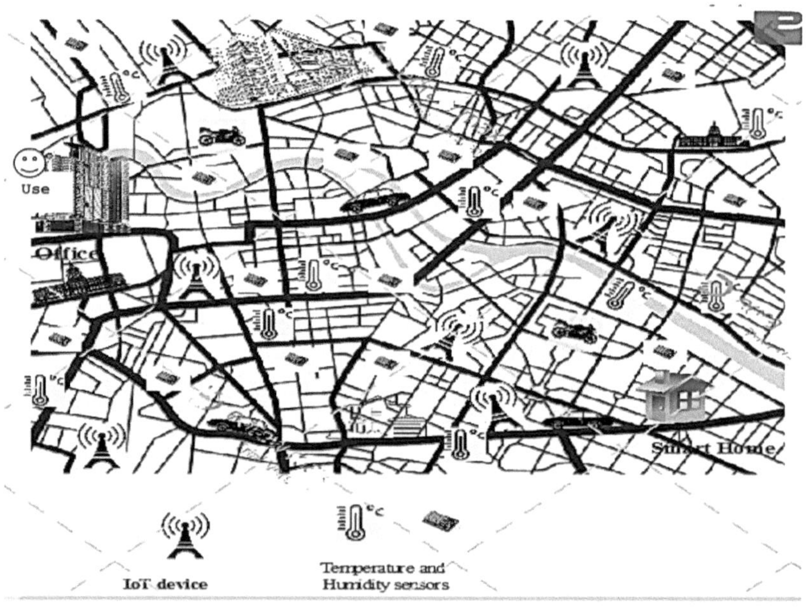

FIGURE 7.5 Intelligent home application scenario using IoT infrastructure in a smart city.

7.4.1 DECISION TREE-BASED LEARNING MODEL

The decision tree-based learning model generates a decision tree based on the attributes. The attributes are considered in the priority order one after the other. In this category, two learning models are considered – C4.5 and RDF:

1. **C4.5:** The C4.5-based learning model is trained using training set and a decision tree is generated. The decision tree formed is used to take decision about classifying unknown inputs. The generated decision tree must be as balanced as possible. For training set S and an item X ∈ S, the number of bits needed to decide whether X is positive or negative is estimated using entropy. If ρ(+) and ρ(−)are percentage of positive and negative examples in S, respectively, the entropy on the training set can be defined using Eqn. (1) [21]:

$$H(S) = -\rho_{(+)} \log_2 \rho_{(+)} - \rho_{(-)} \log_2 \rho_{(-)} bits \tag{1}$$

The gain is measured by obtaining the difference between entropy before the split and after the split. The expected drop in the entropy is obtained using Eqn. (2) [21]:

$$Gain(S,T) = H(S) - \Sigma_{v \in Value\ S(T)} \frac{|S_v}{|S|} H(S_v) \tag{2}$$

where; 'v' is all possible value of 'T' and 'S_v' is the subset for which 'X_T = v.' During decision tree generation attributes are considered in the order of gain in the entropy, i.e., the attribute with the highest gain is given priority over other.

2. **Random Decision Forest (RDF):** It starts by taking a randomly chosen subset of training data S_r and randomly chosen subset d of attributes. λ different decision trees grow in parallel. The information gain is evaluated on S_r instead of full training set. A new data point D is classified using each of the trees S_1,S_2, ..., S_λ and ensemble technique is applied to take the consensus decision [22].

7.4.2 FUNCTION-BASED CLASSIFIERS

The function-based classifiers have a utility function associated with them. The function needs to be optimized for taking decision with high precision and accuracy. Two popular function-based learning models are: gradient descent (GD) and NN based model:

1. **Gradient Descent (GD):** It is known as an iterative algorithm, which starts by choosing a point on a function, randomly. Then step by step it travels down to the slope until the lowest point of the function is reached [23]. The GD algorithm is most suitable when the optimal points cannot be determined by simply equating the slop of the function to zero.
2. **Neural Network (NN):** This classification algorithm is a supervised learning method and is based on back-propagation algorithm. The NN model has an input layer, hidden layer, and output layer. Each of the neurons in the hidden layer transforms weighted input x_i from the previous layer using Eqn. (3) [24]:

$$z = x_1 * \omega_1 + x_2 * \omega_2 + \ldots + x_n * \omega_n \quad (3)$$

For an unknown input y, the predicted output $d = 0g(\bar{y}, \bar{w}, \bar{T})$, where the weight $\bar{w}$ is initialized randomly and adjusted iteratively using error (E) prediction using Eqn. (4) [24]:

$$E = |z - d| \quad (4)$$

The weights are adjusted as per the Eqn. (5) [24]:

$$w_i = w_i + E * m * y_i \quad (5)$$

with 'm' being the learning rate of the NN.

7.4.3 RULE BASE CLASSIFIERS

Rule-based classifiers are based on a list of rules of the form IF C_1 AND C_2 AND C_3 AND ……. THEN class X [25]. The rules are ordered based on the rules quality or based on class. Rule-based classifiers uses the optimized set of rules to take classification decision. Two most efficient and popular rule-based techniques are: non-nested generalized exemplars (NNge) and decision table (DT):

1. **Non-Nested Generalized Exemplars (NNge):** These are the hybrid learners which combine the advantages of induction and instance-based learners. Examples are combined (merge) and most generalized form is kept in the memory. The generalized exemplars represent more than one training example. If there are n attributes in the example from the training set, in geometric terms each of the generalized exemplars represents a geometric figure of n-dimensional covering a finite area. Upon addition of new example in the database, NNge forms a new generalization by joining it to the nearest neighbor of the same class. For classification of new example, in some of the cases rule determines the class and in others, nearest exemplar is used. The nearest exemplar is determined using Euclidean distance ($Euclid_{Ef}$) calculated by the formula in Eqn. (6) [26]:

$$Euclid_{Ef} = W_f \sqrt{\sum_{i=1}^{m} \left(W_i \frac{E_i - f_i}{\max_i - \min_i} \right)^2} \tag{6}$$

   where; E_i *and* f_i are the ith feature value in example and exemplar, respectively with exemplar weight W_f and feature weight W_i.

2. **Decision Table (DT):** It is a learning model is a table of different decisions to be taken based on different conditions. The output of DT is an action set. The DT can also be represented as a decision tree consisting of switch statements or series of if-then-else statements.

It has been observed that one of the learning models used for web service classification would yield the best results. The set of web services misclassified using different classification scheme would not necessarily overlap. Different classification schemes result in complementary information about the classification of web services which could be harnessed to improve the performance of the selected web service classifier. Ensemble technique is useful to improve the classification by combining individual opinion to derive a consensus classification decision. By combining significantly different learning models results into a learning model which take more accurate classification decisions [27]. For an L number of learners resulting from L algorithms, the label output of learners is represented using k-dimensional binary vector $[c_{i,1}, c_{i,2}, \ldots, c_{i,k}] \in \{0,1\}$, for $i = 1, \ldots, L$. The ensemble decision of the class will be resulted from the Ensemble learning model. The Ensemble Learning scheme produces more effective and accurate classification [27]. The important issue in combining the classification techniques is that the scheme is particularly useful if the learners are different [28]. The fact that the chosen learning algorithms are potentially different from one another is assessed in the next section.

7.4.4 CHOOSING THE CLASSIFICATION TECHNIQUE

Based on the past study and performance of learning models in web service classification, initially we selected 10 learning models for evaluation. The learning models considered is GD, NN, NNge, Prism, CART, support vector machine (SVM), DT, C4.5, RDF, and decision stump (DS). 10-fold cross validation-based training and testing of each of the learning model is performed using QWS dataset [29]. The web service classification models are compared on various parameters such as – accuracy, average absolute error (AAE), and average relative error (ARE) along with kappa statistics and visual analysis using model performance chart. Based upon the observation, top two techniques from each of the function based, rule based, and tree-based classifier is selected. The top six learning models considered for further analysis are – GD, NN, NNge, DT, C4.5, and RDF.

The experiments are conducted on a machine with Intel Core i5 CPU 3.0 GHz, Windows 8 platform. The QWS dataset with 364 labeled web services and nine QoS parameters along with WsRF value are used to conduct the experiments. The details of QWS dataset are presented in Table 7.1.

7.4.5 ANALYSIS AND OBSERVATIONS

In order to perform the empirical study, three cases are considered as shown in Table 7.2. In case 1, nine QoS parameters and WsRF score are used to obtain the performance statistics for each of the Six learning models used for web service classification. In case 2, only nine QoS parameters are considered for evaluation of learning models. Whereas in case 3, ensemble-based feature selection technique is applied and the six most relevant parameters RT, TP, SU, RL, and DO are considered for learning models performance evaluation. In each of the three cases, learning models are evaluated using performance parameters as detailed in Table 7.3.

TABLE 7.1 QWS Dataset

Nine QoS Parameters	WsRF	Class
• Throughput (TP); • Best practices (BP); • Susceptibility (SU); • Response time (RT); • Reliability (RL); • Latency (LT); • Availability (AV); • Compliance (CM); and • Documentation (DO)	A web service relevancy function based on cost, TP, Accessibility, RL, AV, and interoperability analysis.	1. Platinum (high); 2. Gold; 3. Silver; and 4. Bronze (low)

TABLE 7.2 Three Cases Considered for Empirical Study

Case	Parameters Considered for Classification
Case 1	Nine QoS parameters + WsRF score
Case 2	Nine QoS parameters without WsRF score
Case 3	Five QoS parameters obtained using feature selection

TABLE 7.3 Performance Evaluation Parameters for Web Service Classification

Parameter	Significance for Web Service Classification
Accuracy	The measure of the quality or state of being correct in the classification of web service.
Kappa statistics	Measure the classifier performance using expected accuracy and the observed accuracy of classification.
Precision (P)	Precision of web service classification is the probability of a positive prediction being correct.
Recall (R)	Measures the proportion of web services belonging to the positive class and were correctly predicted as positive.
F-Measure	F-measure is the harmonic mean of precision and recall and evaluated as: $\text{F-measure} = \frac{2*P*R}{P+R}$
AAE	Average of difference between actual class and web service classification prediction over all web services.
ARE	Average of ratio of difference between actual class and web service classification prediction to actual class for all classification predictions.

In this chapter, we used Cochran's Q test [30, 31]. It is a non-parametric test and provides exact analysis to determine that in a group of *m* learning models, whether one or more learning model differ significantly among themselves. A matrix similar to contingency matrix is used during the Cochran's Q test. Each of the output of the classifier on the testing dataset is noted in termPs of successful classification (1) or failure (0), to any of the four classes (mentioned in Table 7.1). Cochran's Q test for three cases is applied and the test results are summarized in Table 7.4. The mean, standard deviation and frequency count for value *0* and *1* is presented in Table 7.4. In order to apply Cochran's Q test, we need to determine whether the six learning algorithms perform equally or not. Therefore, we need to test two hypotheses H_0 vs H_1. During the test significance value α is set to 0.05, i.e., 95% confidence interval.

TABLE 7.4 Sheskin Multiple Comparison Test

Learning Algorithm	Different (P < 0.05)		
	Case 1 (MRD 3.3507%)	Case 2 (MRD 7.9990%)	Case 3 (MRD 7.5271%)
(1) C4.5	(3) (4)	(3) (4) (6)	(3) (4) (6)
(2) DT	(3) (4)	(3) (4) (6)	(3) (4) (6)
(3) GD	(1) (2) (5) (6)	(1) (2) (5)	(1) (2) (5)
(4) NN	(1) (2) (5) (6)	(1) (2) (5)	(1) (2) (5)
(5) NNge	(3) (4)	(3) (4) (6)	(3) (4) (6)
(6) RDF	(3) (4)	(1) (2) (5)	(1) (2) (5)

Note: The numbers in pairs represents similarity among the learning models. Minimum required difference for each case is mentioned. The significance level α = 0.05 indicate the confidence interval of 95%.

1. **Null Hypotheses (H_0):** All of the web service classification algorithms perform equally.
2. **Alternative Hypothesis (H_1):** At least one of the web service classification algorithm is significantly different from the other learning models which are significantly different are shown along with the minimum required difference (MRD) value, for each case. The six learning models are compared for their accuracy of classification, kappa statistics, and F-measure value. The values are measured in each of the three cases and summarized in Table 7.5. The results of multiple comparison test and comparison based on accuracy and other parameters are summarized as follows.

The web service classification using GD performs differently than C4.5, DT, NNge, and RDF algorithms for web service classification. For the rest of the other algorithms, no significant difference is observed.

Similarly, NNge and RDF are significantly different from one another and also from GD and NN algorithms.

The average accuracy of NNge is better than DT algorithm. Based on kappa statistics and F-measure, NNge perform better than DT in all three cases. RDF outperforms over C4.5 and GD is preferred over NN technique for web service classification. Overall, we observed that three techniques – NNge, RDF, and GD are different from one another, and they perform better than other techniques in their respective categories. Therefore, NNge, RDF, and GD techniques for web service classification are selected for further consideration.

The pairwise comparison of ensemble technique is done with GD, NNge, and RDF based on the accuracy of the prediction parameter. The simultaneous confidence intervals obtained as a result of applying the Dunnett test with the ensemble technique. As per the observation, all three intervals include zero, the performance of each of the NNge, RDF, and GD models is found not significantly different from the ensemble technique. The observations of Dunnett test for simultaneous confidence intervals are summarized as follows:

- The ensemble technique combines the advantages of NNge, RDF, and GD.
- The performance of ensemble technique is comparable with GD, NNge, and RDF web service classification technique.
- Improve in the confidence for taking web service classification decisions.
- Overall, the ensemble technique-based learning model can be used for web service classification.

TABLE 7.5 Comparison of Six Web Service Classification Algorithms on Accuracy, Kappa Statistics, and F-Measure

Performance Measurement Parameters		Rule-Based		Decision Tree-Based		Function-Based	
		DT	NNGe	RDF	C4.5	NN	GD
Accuracy	Case 1	100	99.3217	99.8642	99.8642	92.8061	92.1359
	Case 2	72.5286	72.5286	82.1529	69.6155	82.4187	86.8243
	Case 3	70.1002	73.2527	79.022	71.3297	85.6154	82.6154
Kappa Statistic	Case 1	1	0.9926	0.9963	0.9963	0.9090	0.9054
	Case 2	0.6026	0.6046	0.755	0.5809	0.7423	0.805
	Case 3	0.5814	0.6161	0.6949	0.5900	0.7874	0.7880
F-Measure	Case 1	1.000	0.994	0.998	0.998	0.924	0.933
	Case 2	0.691	0.714	0.820	0.696	0.816	0.858
	Case 3	0.700	0.724	0.783	0.705	0.848	0.848

7.5 DISCUSSION

The empirical study of performance of six web service classification learning models – GD, NN, DT, NNge, C4.5 and RDF is done. The statistical test on the output generated by each of the six models is conducted in three different cases. Based on the Chorchorans Q test, it is found that $p < 0.05$ for each of the three cases, i.e., at least one or more number of learning models used for web service classification differ significantly. In order to determine which algorithm, differ from others, the Sheskin multiple comparison test is conducted. Moreover, the six web service classification techniques are compared on accuracy, kappa statistics and F-measure value. From Sheskin multiple comparison test and performance comparison of six web service classification techniques, we observed that:

i. Learning models GD, NNge, and RDF differ significantly from one another;
ii. GD (function-based), NNge (rule-based) and RDF (decision tree based) perform better than the other learning models in their respective group.

In order to combine the advantages of top three learning models, the ensemble-based technique is used. The ensemble technique is compared with RDF, NNge, and GD model by conducting Dunnett's test. We analyzed that the introduction of the ensemble technique:

- Reduces error rate (AAE and ARE) in predicting the class of web service;
- The performance of the technique is comparable with the top-ranked learning models;
- Improved confidence for taking web service classification decision.

7.6 CONCLUSIONS

A city is internally consisting of many heterogeneous systems. In order to bring smartness to the city to be termed as smart city, all the heterogeneous systems need to work together. The IoT acts as glue to enable heterogeneous systems to work coherently. The necessary infrastructure for implementation of smart city and deployment of applications to access the resources in smart city is facilitated by IoT. In order to realize the IH using IH application, the performance of web services is very important. The IH service internally requires accessing many web services to cooperate and work in cohesion with each other. The use of web service technology as part of IoT architecture has many advantages and is perfectly suitable for implementation of smart applications. The web service technology being at the heart of the operation of smart city needs to be efficient. The efficiency of web service selection operation can be improved by using classification learning models.

An empirical study of six learning models used for web service classification is performed. The Cochran's Q test is applied to determine whether one or more number of learning models differ significantly. Next, the Sheskin's multiple comparison test is applied to determine exactly which technique is different from other. The result of Sheskin's test indicates that three web service classification models – GD, NNge, and RDF differ significantly from one another. Also, using these three learning models, the best result for web service classifications is achieved in terms of reduced error rate, and improved efficiency. Further, the confidence in taking web service classification can be improved by combining classifiers with an ensemble technique. Three classification models GD, NNge, and RDF are combined using an ensemble. Dunnett's test is applied, which advocates that the ensemble performs equivalently as that of three base learning models – GD, NNge, and RDF. Moreover, the reduction in AAE and ARE values is observed, and an improved consensus decision is obtained by using the ensemble technique.

KEYWORDS

- **decision table**
- **information and communications technologies**
- **intelligent home**
- **Internet of things**
- **neural network**
- **quality of life**

REFERENCES

1. Swamy, S. N., Jadhav, D., & Kulkarni, N., (2017). Security threats in the application layer in IoT applications. In: *Proceedings of the International Conference on IoT in Social, Mobile, Analytics and Cloud, I-SMAC 2017* (pp. 477–480). doi: 10.1109/I-SMAC.2017.8058395.
2. Gusmeroli, K., Haller, S., Harrison, M., Kalaboukas, K., Tomasella, M., Vermesan, O., & Wouters, (2009). *Vision and Challenges for Realizing the Internet of Things, 1.*
3. Vision, A., Elements, A., Issues, S., Mpstme, N., & Ram, J., (2017). *Internet of Things (IoT)*, 492–496.
4. Tuwanut, P., & Kraijak, S., (2016). *A Survey on IoT Architectures, Protocols, Applications, Security, Privacy, Real-World Implementation and Future Trends,* 6. doi: 10.1049/cp.2015.0714.
5. Umar, Z., Taj, S., & Adnan, S., (*2017*). CPEC: Security threats and measures. *Proc. Int. Conf. IoT Soc. Mobile, Anal. Cloud, I-SMAC 2017*, 477–480. doi: 10.1109/I-SMAC.2017.8058395.
6. Naru, E. R., Saini, H., & Sharma, M., (2017). A recent review on lightweight cryptography in IoT. *Proc. Int. Conf. IoT Soc. Mobile, Anal. Cloud, I-SMAC, 2017*, 887–890. doi: 10.1109/I-SMAC.2017.8058307.
7. United Nations, (2019). *World Urbanization Prospects: The 2018 Revision*. United Nations: New York, NY, USA. ISBN 978-92-1-148319-2.
8. Scruggs, G., (2018). *Everything we've Heard About Global Urbanization Turns out to be Wrong, 2018,* 10. Reuters.
9. Bowerman, B., Braverman, J., Taylor, J., Todosow, H., & Von Wim-Mersperg, U., (2000). The vision of a smart city. In: *2International Life Extension Technology Workshop* (Vol. 28). Paris.
10. EU, Cities of Tomorrow, (2011). *Challenges, Visions, Ways Forward.* In: European Commission, Directorate General for Regional Policy.
11. Smart City. Available online: http://smartcities.gov.in/content/innerpage/what-is-smart-city.php (accessed on 21 December 2022).
12. Monzon, A., (2015). *Smart Cities Concept and Challenges: Bases for the Assessment of Smart City Projects*. Springer International Publishing Switzerland.
13. Bouk, S. H., Ahmed, S. H., Kim, D., & Song, H., (2017). Named-data-networking-based ITS for smart cities. *IEEE Commun. Mag., 55*, 105–111.
14. Khateeb & Samah, (2018). IoT architecture a gateway for smart cities in the Arab world. In: *2018 15 Learning and Technology Conference, L and T* (pp. 153–160).
15. Silva, B. N., et al., (2018). Urban planning and smart city decision management empowered by real-time data processing using big data analytics. *Sensors (Switzerland), 18*(9), 6–12. doi: 10.3390/s18092994.
16. Corcoran, P., (2016). The internet of things, why now, and what's next. *IEEE Consumer Electronics Magazine, 5*(1), 63–68.
17. Purohit, L., & Kumar, S., (2019). Web services in the IoT and smart cities: A study on web service classification. *IEEE Consumer Electronics Magazine, 8*(2), 39–43.
18. Patru, I. I., Caraba, M., & Barbulescu, M., (2016). Smart home IoT system. In: *IEEE 15 RoEduNet Conference: Networking in Education and Research* (pp. 1–6). Bucharest. doi: 10.1109/RoEduNet.2016.7753232.

19. Mohanty, S. P., Choppali, U., & Kougianos, E., (2016). Everything you wanted to know about smart cities: The internet of things is the backbone. *IEEE Consumer Electronics Magazine, 5*(3), 60–70.
20. 20. Crowley, D. N., Curry, E., & Breslin, J. G., (2016). Citizen actuation for smart environments. *IEEE Consumer Electronics Magazine, 5*(3), 90–94.
21. 21. Mohanty, R., Ravi, V., & Patra, M. R., (2010). Web-Services classification using intelligent techniques. *Expert Systems with Applications, [online]. 37*(7), 5484–5490. Available: http://www.sciencedirect.com/science/article/pii/S0957417410001028 (accessed on 21 December 2022).
22. 22. Hosmer, D. W., & Lemeshow, S., (2000). *Applied Logistic Regression* (pp. 1–43). John Wiley & Sons, Inc, USA.
23. 23. Cong, G., Bhardwaj, O., & Feng, M., (2017). An efficient, distributed stochastic gradient descent algorithm for deep-learning applications. In: *2017 46 International Conference on Parallel Processing (ICPP)* (pp. 11–20). Bristol. doi: 10.1109/ICPP.2017.10.
24. 24. Qamar, U., Niza, R., Bashir, S., & Khan, F. H., (2015). A majority vote-based classifier ensemble for web service classification. *Bus Inf. Syst. Eng., [online], 58*(4), 249–259. Available: http://link.springer.com/article/10.1007/s12599-015-0407-z (accessed on 21 December 2022).
25. Mustafa, A. S., & Swamy, Y. S. K., (2015). Web service classification using multi-Layer perceptron optimized with tabu search. In: *Proc. IACC* (pp. 290–294).
26. Zaharie, D., Perian, L., & Negru, V., (2011). A view inside the classification with non-nested generalized exemplars. In: *Proc.* IADIS'11 (pp. 19–26).
27. Kuncheva, L. I., (2004). *Combining Pattern Classifiers - Methods and Algorithms* (pp. 16–148). John Wiley & Sons, Inc, USA.
28. Kittler, J., Hatef, M., Duin, R. P. W., & Matas, J., (1998). On combining classifiers. *IEEE Trans. on Pattern Analysis and Machine Intelligence, [online], 20*(3), 226–239. Available: http://ieeexplore.ieee.org/stamp/stamp.jsp?arnumber=667881 (accessed on 21 December 2022).
29. Al-Masri, Q. H. M. E., (2007). QoS-based discovery and ranking of WSs. In: *Proc. ICCCN'07* (pp. 529–534).
30. Chen, D., & Cheng, X., (2001). *Statistical Comparisons of Multiple Classifiers.* Deptt. of CSE, UoM, USA, Tech. Rep., No TR 01-025.
31. Schapire, R. E., (2001). Random forests. *Machine Learning, [online], 45*(1), 5–32. Available: http://link.springer.com/article/10.1023/A:1010933404324 (accessed on 21 December 2022).

CHAPTER 8

Applications and Challenges of IoT-Based Smart Homes

P. S. PATHEJA, YATIN KALRA, and AKASH TYAGI

Associate Professor (Sr.), School of Computing Sciences and Engineering, VIT Bhopal, Madhya Pradesh, India

Student, Bachelor of Technology, School of Computing Sciences and Engineering, VIT Bhopal, Madhya Pradesh, India

ABSTRACT

With the invention and implementation of a highly intelligent connected environment, Smart Homes are currently turning into reality. The smart solution has fornicated the world to adopt a lifestyle that magnifies the use of technology in our mundane lives. Home machines and gadgets are interconnected by means of restrictive or standard TCP/IP conventions considering better management and audit. With the expanded implementation of IoT gadgets in a keen home climate, the attacks and challenges likewise should be tended to improve and construct a protected and versatile network safety-based IoT smart setup. Regardless of any type of network, an intelligent home is likewise inclined to security threats and vulnerabilities. This chapter targets introducing smart homes, devices, and the significance of security in the smart home climate.

8.1 INTRODUCTION

With the headway in technology, there have been various methodologies to advance the personal need for individuals. The use of perspicuous approaches

Intelligent Sensor Node-Based Systems: Applications in Engineering and Science, Anamika Ahirwar, Piyush Kumar Shukla, Prashant Kumar Shukla, and Ruby Bhatt (Eds.)

as far as electronic help had been explored and worked upon to expand the life span of humanoid existence. Quite possibly, the most mainstream decisions for this objective have been the development of smart homes [1], where individuals with affliction have been reserved and given special benefits. Among the various kinds of smart homes intended for checking individuals in business and exploration-based conditions, keen homes for older consideration have been a champion. In these IoT Based Smart homes, sensors have been situated in various areas to recognize the exercises and developments in the home. Perhaps the greatest bit of leeway of these homes is the robotized checking and assistance framework, which makes it simpler and more proficient to live.

8.2 IOT-BASED SMART HOMES

IoT is the most widely recognized idea of things that are coherent, controlled, addressed, and located by means of the web in the 21 century. All the things present in our surroundings can be connected to the internet because of the high-paced emerging computing and corresponding competencies. Figure 8.1 shows the structure of a smart home having various IoT-associated devices [1, 6].

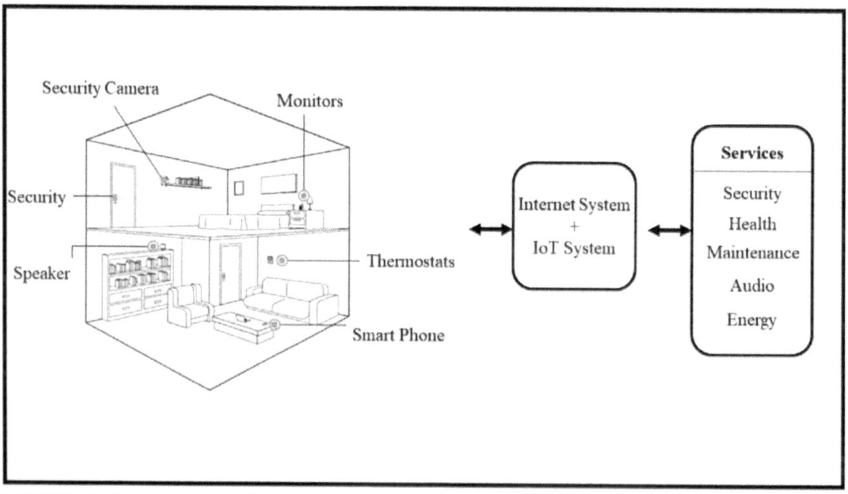

FIGURE 8.1 Structure of a smart home having various IoT-associated devices.

The accompanying sensors are imperative to build up a brilliant home:

1. **Fire Indicator:** Fire alarm is fundamental sensors to deploy a smart home and to shield the home from the harmful effects of

fire. The capacity of the fire indicator is to distinguish the main indication of fire nearly as fast as it can and protect the living souls. It generally makes caution noises to make the souls alert in the house. Figure 8.2 shows an example of a commercial fire indicator.

FIGURE 8.2 Commercial fire and smoke indicators [4].

2. **Humidity Finder:** Humidity sensor is utilized in a smart home to recognize the leakage of water. It can make the living souls in home aware of the same, so they can resolve the issue quickly, staying away from any sort of damage. It can be put around water warmers, fridges, or any other water source, etc. Figure 8.3 shows an example of a commercial humidity finder.

FIGURE 8.3 Commercial humidity finder [3].

3. **Smart Indoor Regulator:** The keen indoor regulator gives command over the warming and cooling in brilliant homes – from any area. The ideal indoor regulators change the temperature on the basis of each and every individual room to keep up the ideal temperature perfect.
4. **Motion Sensors:** Movement sensor perceives development in a space. They can alarm rapidly if there is any dubious development inside the space, or if the doorways or windows has changed their mode to open or close. They can even kill the lights on and relying on the custom settings by client. These sensors work as an additional eye when nobody is at home. They can be called the first line of safety for a keen home.
5. **Security Cameras:** These security cameras permit the organizer to get the positions of various individuals in the home with the help of a smart controller. Whether the proprietor is inside at home or at work and somebody is at the house, the proprietor may get the info and make the quick essential moves. Figure 8.4 shows an example of a commercial Security Camera.

FIGURE 8.4 Commercial security cameras.

6. **Smart Plugs:** A smart plug is a power holder that connects to a conventional power plug and incorporates it into your smart home organization, permitting you to control whatever you plug into it from an application on your cell phone or with your voice through a remote helper (Figure 8.5).

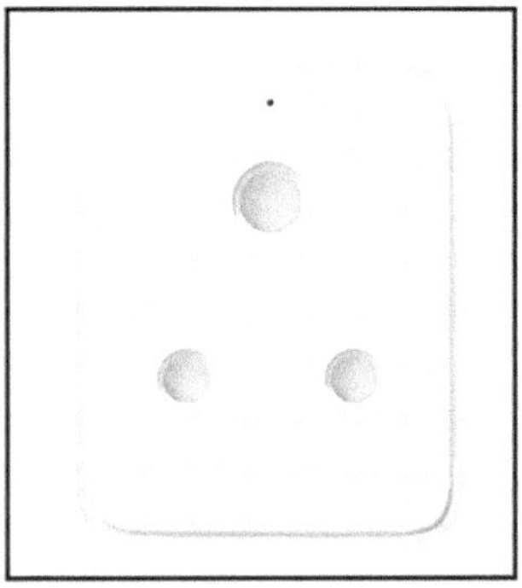

FIGURE 8.5 Smart plugs.

7. **Smart Security Bell:** Web-associated doorbell that advises the cell phone or other electronic gadget of the mortgage holder when a guest shows up at the entryway. It actuates when the guest presses the catch of the doorbell, or then again, when the doorbell detects a guest with its underlying movement sensors. The keen doorbell lets the property holder utilize a cell phone application to watch and chat with the guest by utilizing the doorbell's implicit top-quality infrared camera and receiver. They can be either battery-worked or wired. Some brilliant doorbells likewise permit the client to open the entryway distantly by utilizing a shrewd lock (Figure 8.6).

FIGURE 8.6 Smart security bell.

8.3 ARCHITECTURE OF SMART HOMES

Architecture system of IoT smart homes is depicted in Figure 8.7.

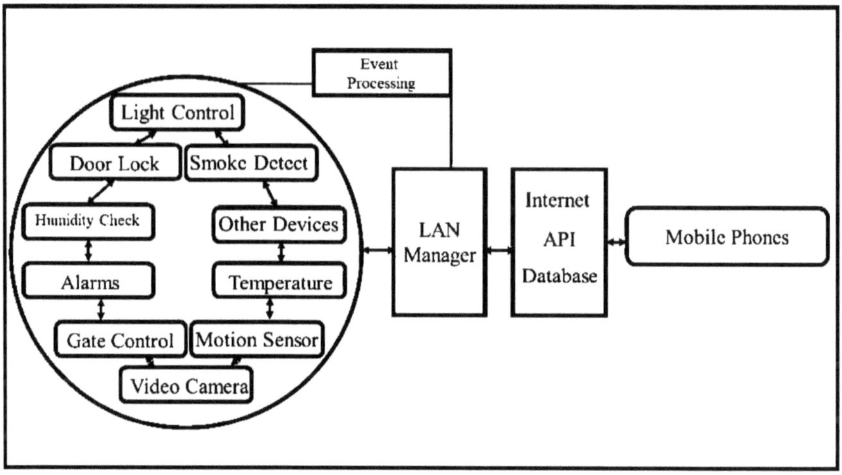

FIGURE 8.7 Architecture of IoT-based smart homes.

Highlights of architecture of smart homes are:

i. Sensors are considered as the primary building blocks or the base of this technology. Sensors are the devices which are used to detect and measure various physical property and records and develop mechanisms to generate responses of various types. Smart homes use a combination of various types of sensors in order to accomplish a smooth functioning of events [2].
ii. Processors are used to manipulate, transfer, stimulate, and manage the information collected by sensors with other components of IoT Smart Homes system.
iii. Segment of programming as APIs, permitting outside applications to execute it, given it follows the pre-characterized boundaries design. Such an API can deal with sensors information or essential activities.
iv. Actuators are to arrange and to process devices. It understands and interprets the command to perform the necessary action to the order language structure that the gadget can execute.

v. Database is used to record the confidential information fetched from the sensors. It can also be used for investigation, information introduction, and visualization.

8.4 BENEFITS OF INTERNET OF THINGS (IOT) IN SMART HOME

1. **Control and Monitoring:** With the use of IoT solutions in the smart home people can get extensive control over the entire household. This ranges from switching off and on various household appliances to controlling the security of the home by a simple web or mobile application.

   All the connected devices of the system collect, analyze, and process data to provide meaningful output to the end-user. A person can now monitor their spending on utilities, energy consumption, track humidity, air quality, and even any motion in the house in his absence.

2. **Optimization of Spending:** The use of IoT-based devices not only provides control to the owner but also enables transparency to the household. As with the help of smart home technology, one can keep track of its spending and wastage, this data can be analyzed to reduce waste and thus helps in optimizing spending.

3. **Environmental Impact:** IoT devices and their various applications are working as a driving force for green energy initiatives. The use of IoT is also helpful in reducing our carbon footprint. The use of IoT and smart grid decreases carbon emission which results in a cut down on pollution.

4. **Improved Comfort:** Home automation is one of the basic implementations of the smart home. Intelligent algorithms not only help in providing a certain level of automation but as well as equipped the device with the capability of making an efficient decision.

   Automatic window blind adjustment, smart refrigerator which checks the expiration dates of the items, IoT-based camera, and detectors have made our livers easier and comfortable.

8.5 TOPOLOGIES OF NETWORK DEVICES IN IOT SYSTEM

Network topology is the setup of various network devices of a communication network. It can be used to portray the logical and physical setup of different kinds of telecom organizations, printers, control radio networks, and PC organizations.

Topology on the internet of things (IoT) means the arrangement by which sensors, actuators, and doors transfer data to each other. There are a number of basic geographies, namely point-to-point, star, bus, and mesh.

8.5.1 POINT-TO-POINT TOPOLOGY

Point-to-point network sets up an immediate association between two network points, as depicted in Figure 8.8.

Transfer of data can occur just between these two hubs, or gadgets. For Example, Bluetooth interface between a cell phone and an earpiece.

The benefit of this topology is that it is a lot more straightforward than mesh or star as it passes the information either in one or many directions from one to another network point.

The impediment is that point-to-point networks are not helpful for IoT as there are too many sensors and points to communicate in the network system.

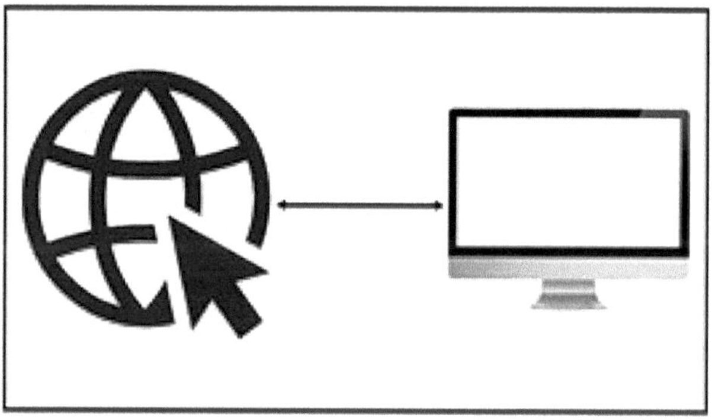

FIGURE 8.8 Point-to-point topology.

8.5.2 STAR TOPOLOGY

In this topology, each connection is associated with a central common hub via point-to-point connection, with each PC by implication associated with all other nodes with the assistance of the center as illustrated in Figure 8.9. All traffic that crosses the network goes through the central hub.

The benefit of this topology is that all the unpredictability in the organization is headed to a central hub, so all different hubs just require to convey in their time or slot.

The impediment is that the link between the central hub and the end node can be very long.

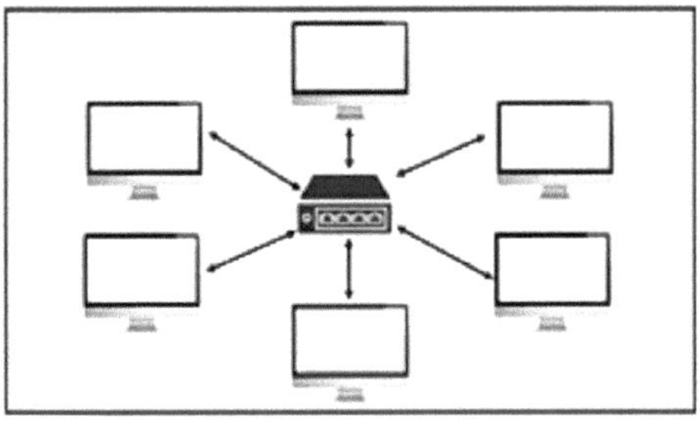

FIGURE 8.9 Star topology.

8.5.3 BUS TOPOLOGY

It is an arrangement of all nodes that are connected with a single cable. The cable to which the nodes connect is called a "backbone," as displayed in Figure 8.10.

The benefit of this topology is that it works very efficiently when there is a small network. Usually, the length of the cable required is less than a star topology. The impediment is that if the common bus breaks the entire segment will fail.

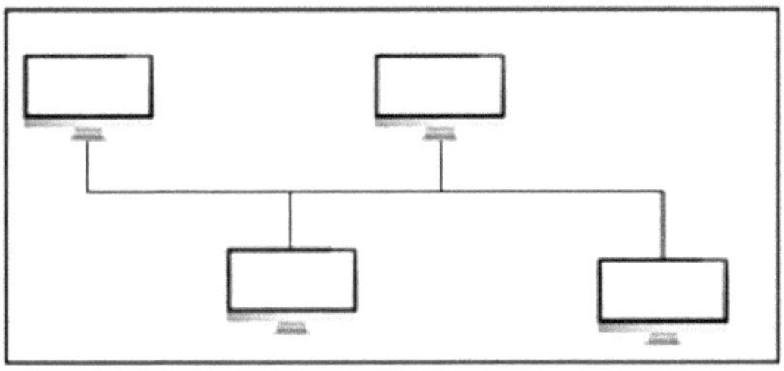

FIGURE 8.10 Bus topology.

8.5.4 MESH TOPOLOGY

This topology is an arrangement in which each network device is connected to each other. This topology ensures proper stimulation of data even if any connection turns down as exhibited in Figure 8.11. This topology is ordinally used for wireless networks.

The benefit of this topology is that it allows distributed transmission even if one node is down. The impediment is that the node has to work as both, i.e., sender and router.

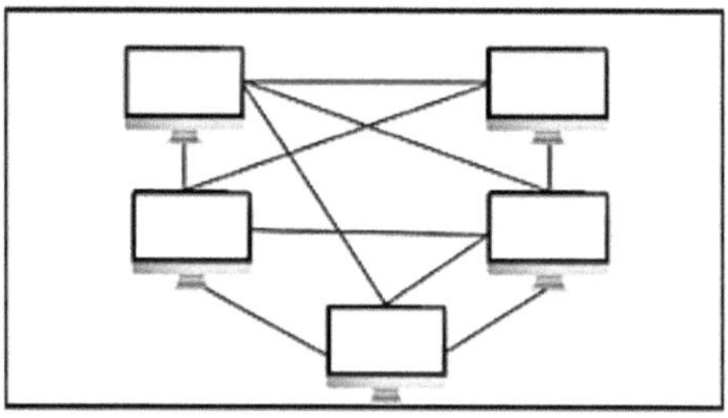

FIGURE 8.11 Mesh topology.

8.6 PROTOCOLS TO TRANSFER DATA IN IOT SYSTEM

Protocols play a vital role in the IoT Technology. IoT protocols empower the system to exchange information in an organized and important manner. Out of these moved bits of information, valuable data can be used by the end client. Because of IoT Protocols, the entire arrangement turns out to be monetarily productive, particularly regarding IoT gadgets [7].

When discussing the IoT, we generally consider the transfer of data. Efficient communication between sensors, gadgets, entryways, workers, and applications is the fundamental trademark that marks the ability of IoT.

8.6.1 CONSTRAINED APPLICATION PROTOCOL (COAP)

As the current internet framework is available for any IoT Device without any restriction. It regularly demonstrates excessively hefty and power-devouring for most IoT devices. It was intended to translate the HTTP model with the goal that it very well may be utilized in prohibitive gadget and network framework (Figure 8.12).

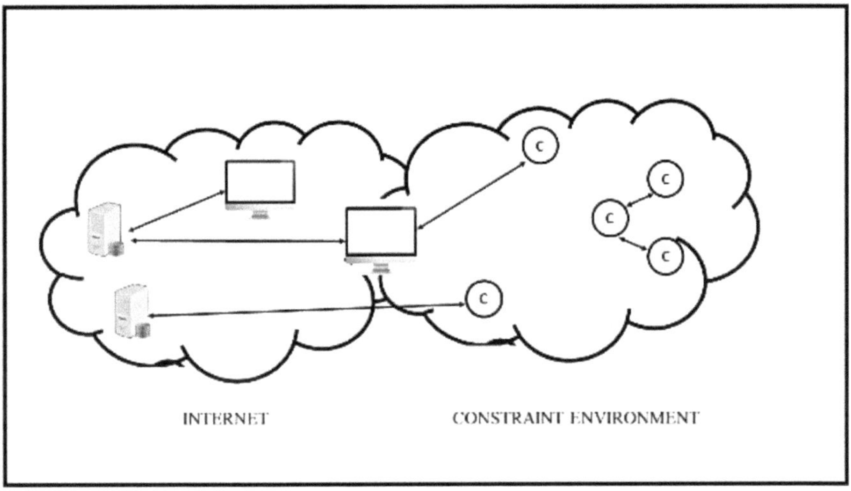

FIGURE 8.12 Constrained application protocol (CoAP).

8.6.2 MESSAGE QUEUING TELEMETRY TRANSPORT (MQTT)

Message queuing telemetry transport (MQTT) is a portable distribution protocol which is commonly used in IoT. It is made for gadgets which uses battery as their power source, MQTT's architecture is simple and portable, giving low power utilization to gadgets (Figure 8.13).

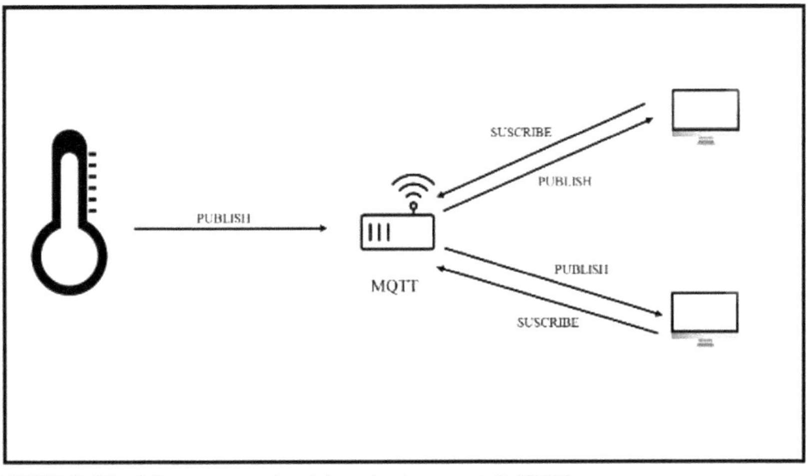

FIGURE 8.13 Message queuing telemetry transport (MQTT).

Having discovered wide application in such IoT gadgets as vehicles, locators, and modern or clean hardware, MQTT reacts well to the accompanying necessities:

- Good reliability;
- Least use of bandwidth.

8.6.3 ZIGBEE

ZigBee protocols are portrayed by low throughputs, low power consumption, and network scope of 100 meters between points. Common applications incorporate sensor organizations, WPAN, home computerization, alert frameworks, and observing frameworks (Figure 8.14).

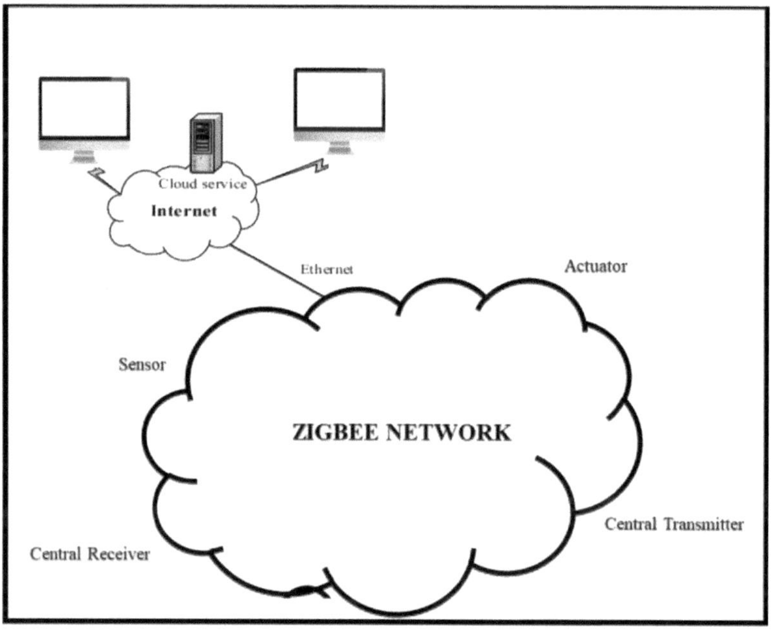

FIGURE 8.14 ZigBee protocol.

8.6.4 BLUETOOTH

Bluetooth technology has privileged wireless connections of various electronic devices, such as personal computer, mouse, video camera, keyboard, alarm systems, keyboard, lighting points, printer, headset, or speakerphone, and more (Figure 8.15).

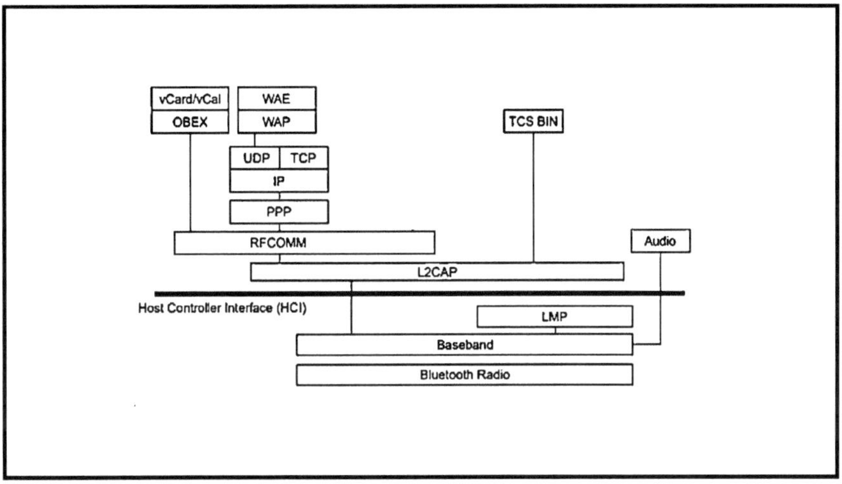

FIGURE 8.15 Bluetooth protocol.

8.6.5 ADVANCED MESSAGE QUEUING PROTOCOL (AMQP)

AMQP is an open-source protocol launched in 2003 has its underlying foundations in the financial sector. It has acquired some ground inside the information transfer technology, its utilization is still very restricted in the IoT. It enables encrypted and secure message transfer between organizations and applications (Figure 8.16).

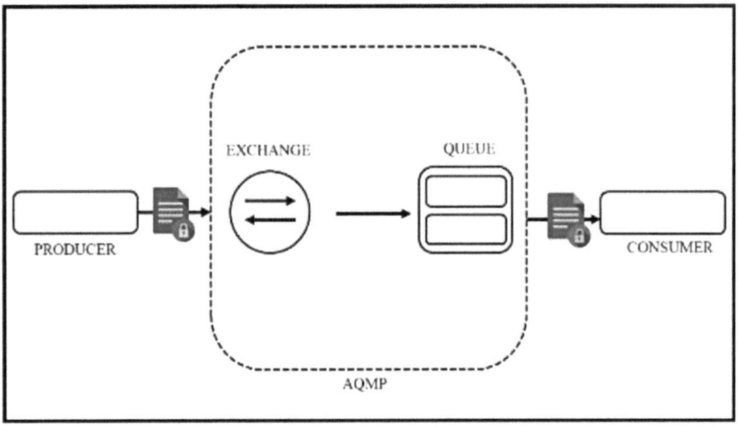

FIGURE 8.16 Advanced message queuing protocol (AMQP).

8.7 CHALLENGES TO IOT-BASED SMART HOMES

The advancement in the field of sensor and IoT technology have provided many smart home solutions, but there are many questions still unanswered. Some of the challenges of IoT Smart homes are listed in subsections.

8.7.1 COST

The use of advanced technology in Smart Homes has augmented the price of IoT-enabled system to an extent that is not affordable for a large section of people. The high price became one of the primary factors behind hostile behavior in the market.

8.7.2 ADAPTIVENESS

The people in most parts of the world are still unaware of these types of technology. People often find it difficult to adjust with technology and new lifestyles. Humongous efforts are to be placed in order to connect people with the technology.

8.7.3 CUSTOMER EXPECTATIONS

When customer expectations for the product do not match with reality at that point, the outcomes can be system disappointments, stranded advances, and lost profitability. Earning customers trust and providing satisfactory outputs to them is a big challenge for the IoT services provider.

8.7.4 SECURITY

Smart Homes uses multiple sensors which are connected to each other by various means. Sensors transfer data from one to another in order to stimulate a response but wherever there is transfer of data or connection there also exist potential safety threats which can compromise the safety of the entire machinery.

Cybercriminals can just cunning their way in through one of the various security weaknesses that are tracked down all through the IoT which can achieve infiltrate of private data. Various IoT contraptions have default passwords which are left unaltered or unpatched issues which can be a threat to the security of the entire structure. Providing a safe network and machinery is a key challenge for all the service providers.

8.7.5 VULNERABLE SMART APPLIANCES

Other than the routers present in the entire system, all the connected appliances are at risk when we talk about cyber-attacks. Hackers use various platforms to find firmware vulnerabilities in the manufactured smart devices and then these vulnerabilities are exploited in order to gain access to the network.

8.8 CYBER SECURITY IN IOT-BASED SMART HOMES

The smart homes sensors depend on human attention to work. IoT sensors gather, transfer, investigate, and follow up on data, offering the latest ways for innovation [3, 5].

In case, this additionally makes chances for confidential data to be compromised. Not only is more data being shared through the IoT, among much more individuals, however more sensitive and confidential information is being shared. Therefore, the cyber threat is exponentially greater.

For example, in IoT based smart homes, one of the major used sensors is Thermostat – a device to control the temperature of a house. Suppose if a cyber attacker however becomes able to hack and take control of the IoT system, the attacker can increase or decrease the temperature to that extent which can further prove to be very harmful for human beings living inside, it can even cause death of persons living inside.

8.9 CYBER ATTACKS FOR IOT-BASED SMART HOMES

Some types of cyber-attacks dedicated for IoT-based smart homes are discussed in subsections.

8.9.1 BOTNETS

Botnets join numerous frameworks to take control of devices from location far away. From here, cyber attackers can fetch classified information and further can proceed with different cyber-attacks. Smart Home Devices are highly vulnerable to these attacks.

"The Mirai botnet, for instance, has influenced 2.5 million gadgets, including smart cameras, switches, and printers. Furthermore, it's just deteriorating. In light of the accomplishment of these attacks, cybercriminals made significantly more developed IoT botnets" (Figure 8.17).

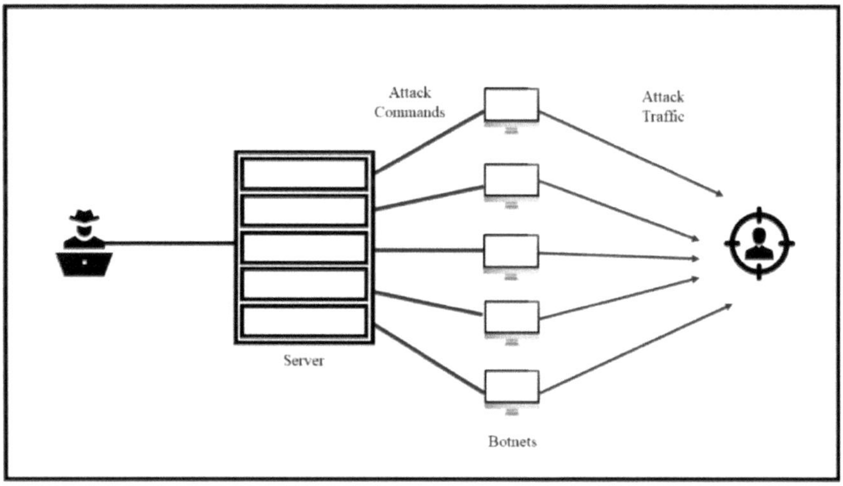

FIGURE 8.17 Botnets cyber-attacks.

8.9.2 MAN-IN-THE-MIDDLE ATTACKS

During the man-in-the-middle attacks, attackers capture communications by attacking communication carrier. They deal with communication to send confidential information. It is a dangerous attack because it is one where the attacker poses as the original sender.

Since IoT gadgets share information progressively, Man-in-the-Middle attacks compromise refrigerators, modern gear, and self-governing vehicles. Their dependence on this action can have terrible outcomes (Figure 8.18).

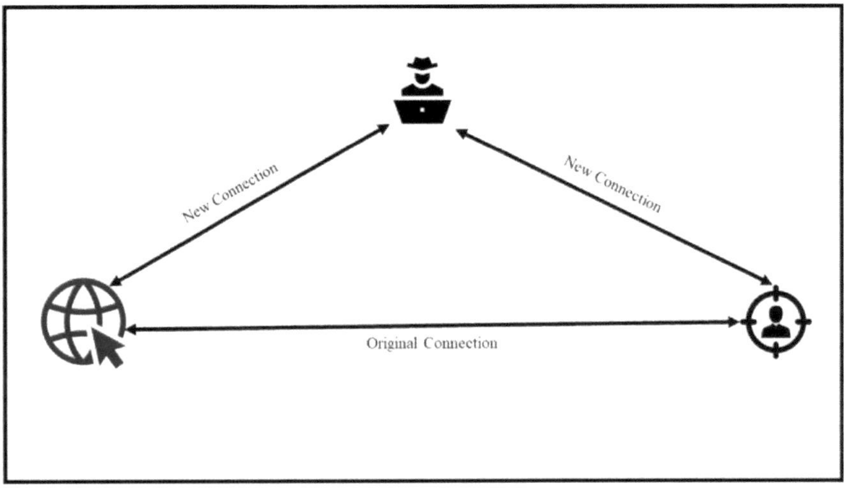

FIGURE 8.18 Man in the middle cyber-attack.

8.9.3 DATA THEFT

Consistently, there are various attacks of data theft. Attackers take control over the information of millions. Cybercriminals now target IoT gadgets, including smartwatches and brilliant indoor regulators, for a similar explanation. It causes them to acquire data about individual clients and associations.

8.9.4 REMOTE RECORDING

There are enough vulnerabilities in IoT gadgets that attackers use to get video or sound film of a target. While these attacks are more uncommon, they are likewise perilous.

It puts organizations at danger of their classified data getting leaked. Regardless of whether an IoT camera is secure, other IoT gadgets with lower security conventions can give attackers what they need to penetrate an organization. At that point they can get to a similar camera not long after (Figure 8.19).

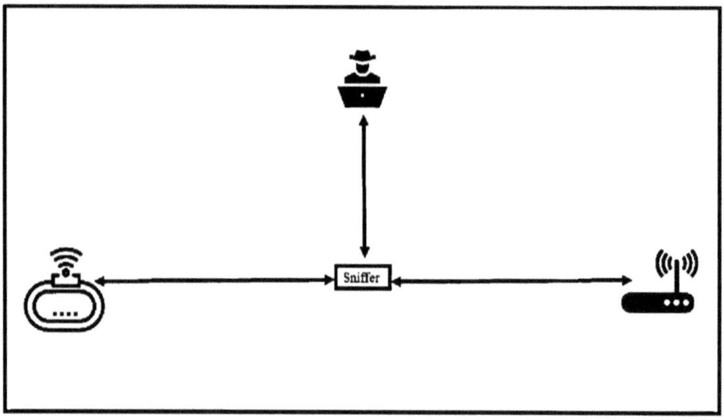

FIGURE 8.19 Remote recording cyber-attacks.

Regardless of these dangers, the fame of IoT gadgets won't quit developing over the course of the following not many years. Yet, it takes effort for producers to get up to speed and foundation the correct security upgrades.

8.10 SECURITY SOLUTIONS FOR IOT-BASED SMART HOMES

There are a few simple advances the two organizations and individual clients can take to make secure their IoT gadgets. Connection with the internet for each IoT gadget encrypted is fundamental. Also, one can do it in a similar way they encrypt their PC association—utilizing a virtual private organization (VPN). It makes an encoded tunnel between the gadget and the web. The association of a gadget goes through it, remaining garbled to any sneaking around party on the network. It's the simplest method to make a private, secure, and safe association. Following are some more steps that can be taken to make IoT Based Smart Homes safe.

8.10.1 SECRET PASSWORD

It is significant that change passwords on PC's, cell phones, and other IoT Devices. One should guarantee that every gadget has a different secret password. Each secret password should be changed frequently. Complex password must be of minimum 16 characters, including upper and lower-case alphabets, numbers, symbols, and different combinations.

Example of secure password – Op)9&66124!mk00+. According to howsecureismypassword.net this password would take 1 trillion years to get cracked.

8.10.2 DISABLE UNIVERSAL PLUG AND PLAY FEATURES

Disable universal plug-and-play features to new highlights. With this component, various gadgets can discover each other and associate with one another automatically. This highlight makes the gadgets more advantageous in light of the fact that one doesn't have to arrange every one of these gadgets independently. UPnP conventions utilize neighborhood networks for interfacing and are subsequently vulnerable against outside attack. In the case of a cyber-attack, outside elements may have the option to access various gadgets all the while. Henceforth, it is a decent practice to turn off the UPnP feature on each gadget.

8.10.3 SEPARATE THE CORPORATE NETWORK

At whatever point conceivable, separate the corporate organization from unmanaged IoT gadgets. This may incorporate surveillance cameras, temperature control gadgets, electronic signage, keen TVs, media focuses, security DVRs and NVRs, network-associated tickers, and organization associated lighting. Use VLANs to separate and monitor different IoT gadgets on the organization and to oversee significant capacities, for example, office tasks, security activities, and clinical hardware. Ultimately, apply an Access Control List, or ACL, to VLANs or organize access ports at whatever point conceivable to restrict correspondence to the least sum that is needed for gadget activity.

8.11 FUTURE OF SMART HOMES

The recent years have witnessed many developments pertaining to Smart homes and IoT. With the advancement in technology, many daily chores are accomplished by the smart and effective implementations of sensors. This revolutionary concept provides numerous options to explore [8].

With technology fueled with power over Ethernet and IoT, the amalgamation of smart homes technology with artificial intelligence (AI) can be predicted as the future technology that will change the world completely.

One of the future aspects of Smart homes can be in the healthcare sector where smart home technology can be used to collect data of the patients which can be further analyzed and can be shared with the concerned medical staff and family in case of any health risk. This will not only save millions of lives but will also ease the burden of the healthcare industry.

Another future of smart homes can be smart cities. The application of smart homes infrastructure on a large scale can accomplish this goal. Smart cities will not only be a comfort for the people but can also be an important development which can be used in various sectors to achieve multiple goals like lesser pollution or automated systems.

Some of the future implementation of smart homes is given below:

1. **Robo Chef:** For people who are not much into cooking, a robot chef is the solution. This machine will be capable of not only preparing food, but it will also be able to modify from its basic algorithms depending on the need of the user.
2. **Smart Fridge:** This is another future machine which will save us a good amount. From lesser energy consumption to recommending items to buy depending upon its availability inside it can be a life changer.
3. **Smart Cameras:** Security is always the topmost priority of any customer. Smart camera is the solution of all security problems. Integrated with motion sensors and facial recognition, alarm, and notification mechanism, this system will be your on-duty guard.

8.12 CONCLUSION AND RESULT

The home advancements using IoT has been exhibited to work wonderful by interfacing the unpredictable smart gadgets to it and the machines were adequately controlled distantly through web. The arranged structure not simply screens the sensor data, like temperature, gas, light, development, yet what's more actuates a communication according to the need, for example, turning on the fan when it gets hot. It furthermore stores the sensor data in the cloud in an advantageous way. This will help the customer with separating the condition of various data in the home at whatever point.

KEYWORDS

- **advanced message queuing protocol**
- **botnets**
- **constrained application protocol**
- **Internet of things**
- **message queuing telemetry transport**
- **virtual private organization**

REFERENCES

1. Bhat, O., Bhat, S., & Gokhale, P., (2017). *Implementation of IoT in Smart Homes, 6*, 149–154. 10.17148/IJARCCE.2017.61229.
2. Gaikwad, P. P., Gabhane, J. P., & Golait, S. S., (2015). A survey based on smart homes system using internet-of- things. In: *2015 International Conference on Computation of Power, Energy, Information and Communication (ICCPEIC)* (pp. 0330–0335). Melmaruvathur, India. doi: 10.1109/ICCPEIC.2015.7259486.
3. Yatin, K., Saket, U., & Patheja, P. S., (2020). Advancements in cyber-attacks and security. *International Journal of Innovative Technology and Exploring Engineering*. doi: https://www.doi.org/10.35940/ijitee.D1678.029420.
4. Patheja, P., Waoo, A., & Nagwanshi, S., (2012). *Current Trends and Research Issues in Bluetooth Communication, 2*(2), 1784–1792.
5. https://www.forcepoint.com/cyber-edu/iot-cybersecurity (accessed on 21 December 2022).
6. Nag, A., Alahi, M., & Afsarimanesh, N., Prabhu, S., & Mukhopadhyay, S. C., (2019). *IoT for Smart Homes*. doi: 10.1049/PBCE122E_ch7.
7. https://behrtech.com/blog/top-10-iot-sensor-types/ (accessed on 21 December 2022).
8. https://www.digiteum.com/smart-home-trends/ (accessed on 21 December 2022).

CHAPTER 9

Intelligent Security System Based on the Internet of Things (IoT)

POOJA GUPTA, SUNITA VARMA, NEERAJ ARYA, and RITESH BHAGEL

Assistant Professor, Department of Information Technology, Shri G. S. Institute of Technology and Science, Indore, Madhya Pradesh, India

Professor and Head, Department of Information Technology, Shri G. S. Institute of Technology and Science, Indore, Madhya Pradesh, India

Student Department of Information Technology, Shri G. S. Institute of Technology and Science, Indore, Madhya Pradesh, India

ABSTRACT

Nowadays, security has become a significant issue, as the COVID-19 epidemic makes everyone feel insecure; everyone wants to protect themselves with their financial assets in their homes and elsewhere. Homes and banks are places where we store our profits and our capital. In any case, we can never be sure of the safety of that benefit behind us. For the most part, we lock the houses when we leave the house. In any case, just closing the house is not enough; there should be a framework that looks after the exercise and reports to the owner accordingly, and operates at the owner's response. The owner wants to have a reliable system where there is no need for so much communication and security. This paper proposes an intelligent security system based on IoT along with a face detection

Intelligent Sensor Node-Based Systems: Applications in Engineering and Science, Anamika Ahirwar, Piyush Kumar Shukla, Prashant Kumar Shukla, and Ruby Bhatt (Eds.)

technique and checking of its liveness with a zooming option from any place. A completely independent, IoT-based system developed for network communication. The developed system is processed by Raspberry pi along with different sensors and Raspberry IP cameras using Python as a programming language. This framework is battery worked if there should arise an occurrence of intensity disappointment. Besides, the house/office proprietor can monitor action occurring in the house utilizing android and web applications associated with the raspberry pi using the web.

9.1 INTRODUCTION

In the present scenario, security and safety have become of utmost importance. The progress in security systems has been immense due to the use of technology. Therefore, various security systems for houses, and offices are present in the market. There always have been efforts to make security systems more secure and robust with the use of the latest technologies so that minimum human intervention and action are required. With smart cameras and many sensors, the intelligent automatic security system may be configured, and the use of these sensors determines the functionality of such sensors [1]. The Wi-Fi feature in security systems helps the user to manage the device globally and monitor it more easily. In the coming few years, it is expected that the latest IoT goods and services will expand exponentially. IoT incorporates a variety of devices and a wide range of connectivity layer technologies. IoT supports easy access to different data sets. The Raspberry Pi is a lightweight, compact computer that provides web communications and signature and good processing power. This architecture primarily includes monitor, real-time data processing providing the surveillance program [2]. This introduced system is very efficient and cost-effective. Compared to existing systems, the device is robust and uses relatively little power. In this project, an intelligent security system is proposed based on face recognition technology with the liveness test. The central processing unit is Raspberry and python programming using deep learning for the implementation of face recognition and the liveness test of users [3].

9.2 LITERATURE SURVEY

Assortments of other most recent advancements accessible are RFID card advances, and biometric ensured frameworks, OTP based, cryptography-based, and some more. Every framework is pertinent for various application zones relying on their innovation utilization. Likewise, there are frameworks that utilization a portion of this security method, yet they don't give a total security framework as there is just single-factor verification.

- P. Bhatia et al. build up a productive face acknowledgment framework for home security framework in which it utilizes histogram of oriented gradients (HOG) for face discovery and local binary pattern histograms (LBPH) calculation for face acknowledgment to perform extremely effective face acknowledgment for security issue [1].
- P. Gupta et al. developed a system, which can be used in homes and societies such that the electrical gadgets and switches can be distantly controlled and observed with or without an android-based application. The system utilizes numerous sensors for following and keeping up the security of your home [2].
- A. K. Gupta et al. talk about vitality-sparing electrical gadget Surveillance and Control framework dependent on IoT. It utilizes two models for little territory or bound premises IEEE 802.11 wireless technology is used while the rest of the appliance of the home is connected with a Wi-Fi network. A street appliance that is connected in one direction is wired connected [3].
- Vadivukarasi et al. "Home security using IoT" proposes in their paper a system in which the door of the house is opened if face recognition is successful. Their system uses Raspberry Pi and a USB camera [4].
- Adiono et al. design an entryway lock system that needn't bother with manual contribution from the client for comfort reasons while additionally staying secure. The framework essentially comprises the STM32L100 micro-controller as its center [5].
- Malche et al. propose a Smart Home and its applications and presents a FLIP design with usage of Smart Home administrations utilizing FLIP through a proposed framework. The proposed framework introduced in this chapter is utilized for observing and controlling smart home conditions [6].
- Chifor et al. propose a lightweight approval stack for smart home IoT applications, where a Cloud-associated gadget transfers input orders to a client's smart mobile phone for approval. This architecture is client gadget driven and addresses security issues with regards to an untrusted cloud platform [7].

- Mohan et al. explains in their paper "PhoneControlled Security Lock Systems" a framework for accessing the door lock using an android app and Bluetooth wireless network and describe all potential scenarios when accessing the door lock [8].
- Kodali et al. focus on home automation by utilizing a similar arrangement of sensors. The influence got by leaning toward this system over the comparative sorts of existing systems is that the alarms and the status sent by the Wi-Fi associated microcontroller managed system can be gotten by the client on his telephone from any distance regardless of whether his cell phone is associated with the web [9].
- Patchava et al. portray a system utilizing a Raspberry Pi module with Computer Vision strategies. Utilizing this, we can control home appliances associated through a monitor-based web [10].

9.3 PROPOSED SYSTEM

In this work, we propose an Intelligent Security System where info pictures are fed to the system through an IP camera and different sensors as indicated by their characterized task. In this project, Raspberry Pi and different sensors are used using the internet of things (IoT) concepts. The project includes a database created from multiple images (40) of each house/office member. The system opens the door of the house/office if the person is authorized by the owner or administrative manager of the home or office, i.e., registered in the database. In-depth learning concepts are used here in the face-to-face process and life, and online services are used to send images to the admin to ensure user authority and are widely proposed to be divided into three units: input unit, processing unit, and application unit.

Figure 9.1 shows the completed unit of the proposed system such that the first unit generates data as it takes input from all sensors, the camera passes, then passes it to a raspberry pi. Then this data moves to the processing unit concerning the database and takes appropriate action on the application layer.

9.3.1 HARDWARE INFORMATION

Raspberry pi is the main processing unit of the hardware system. Ultrasonic sensors are used in the system to detect the presence of a person in front of

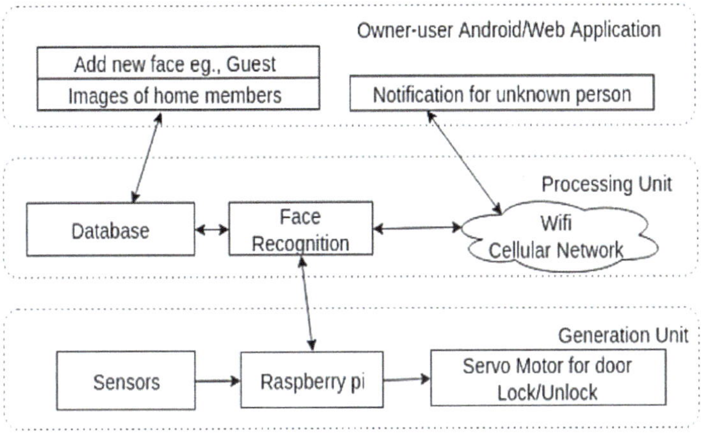

FIGURE 9.1 Proposed system.

a door, then the response to the presence of a human IP camera is activated to capture a visual image. The servo motor is used to lock and unlocks the door; the buzzer is used for the alert facility. Some other sensors are also used to find the smoke and LPG leakage for other detection. Finally, wires and boards are used to configure it, i.e., jumper wires, SD card, led lights, resistors, breadboard, pushbuttons, and 6 V power supply are used in it. Figure 9.2 shows the hardware used in this project.

FIGURE 9.2 Complete hardware.

9.3.2 MAIN MODULES

The main modules for the system are:

1. **Make Final Frames:** This module starts the IP camera and captures images until five images containing faces are not captured. Five images are captured to ensure the robustness of the system. In face recognition, the results with the highest numbers of names, i.e., the names which appear at least three times, are returned.
2. **Liveness Detection Module:** This module does liveness detection on frames that were input from making the final frames module and returns the result of liveness detection.
3. **Face Recognition Module:** This module does face recognition if the liveness detection result is "real." The face recognition result is returned.
4. **Mail Generation Module:** This module generates mail with the attached picture of the new guest.
5. **Text Message Generation Module:** This module generates a message to the homeowner with a message according to the results of liveness detection and faces recognition results.
6. **Receive Mail Response:** This module receives mail response from the homeowner in the case guest at the door was not recognized, and the guest's picture was sent to the homeowner for asking permission to open the door.
7. **Door Unlock Module:** This module opens the door whenever the response received from the homeowner is "yes."
8. **LPG Leakage Module:** This module gets activate if the LPG detector sensor found any leakage information from the sensor, then it starts and sends a signal to the processing unit for alarming and generates a message, mail to owner or admin.
9. **Smoke Detector Module:** This module gets activated if any smoke is detected by the sensor and information send to processing for further message and alarming.
10. **Electronic Appliances Status Module:** This module sends the status of all electronic appliances connected with electricity in

such a way that at evening or night, we get all electronic component status in the form of on/off. At the same time, we can operate it from outside of the office and home also.

9.4 ARCHITECTURAL DESIGN

Figure 9.3 shows the connectivity of all main module of this security system.

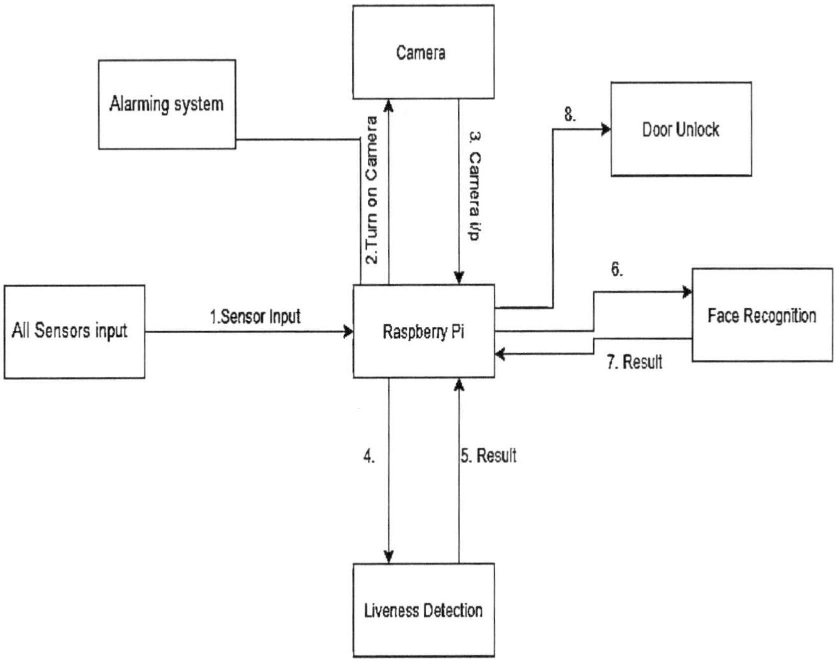

FIGURE 9.3 Architectural design.

9.5 METHODOLOGY

1. As any person comes near to the door, the ultrasonic sensor (which is installed on the door along with the IP camera) senses the person and sends the input signal to the raspberry pi.
2. This input leads to the automatic turning on of IP camera.

3. The IP camera, upon turning on, takes 5 pictures. This picture is input to the raspberry pi.
4. Then raspberry pi processes this picture and starts liveness detection on the faces found in pictures.
5. The result of the liveness detection is fed for further processing.
6. If the liveness detection was successful (i.e., the face in the picture was found real), then Face recognition starts.
7. The result of the face recognition is fed into the algorithm for further processing.
8. If the face recognition was successful, that is the person in the picture is found to be a house member; then the door is opened.
9. LPG detector sensor installed in the kitchen whenever it gets LPG gas as input signal it generates an active signal for raspberry pi, raspberry pi processes signal and activate the alarm with mail message to admin/owner.
10. A smoke detector is connected with raspberry pi as it senses the smoke signal appropriate action done by raspberry pi as define in programming code for alarming and messaging.
11. The raspberry pi system is connected to all electronic devices in such a way that in a specified time it detects all the electronic devices and sends their status to the controller/owner and the controller/owner can also work anywhere.

Figure 9.4 shows that how the security system performs the face detection process and recognition process and taking appropriate action concerning an unknown or known person comes in front of the camera of office/home.

9.6 RESULT

9.6.1 FACE RECOGNITION

Figure 9.5 shows the result of the face recognition process. After the face recognition system can determine it as a known person, and it also shows his/her name on an image by covering a rectangular box on the face.

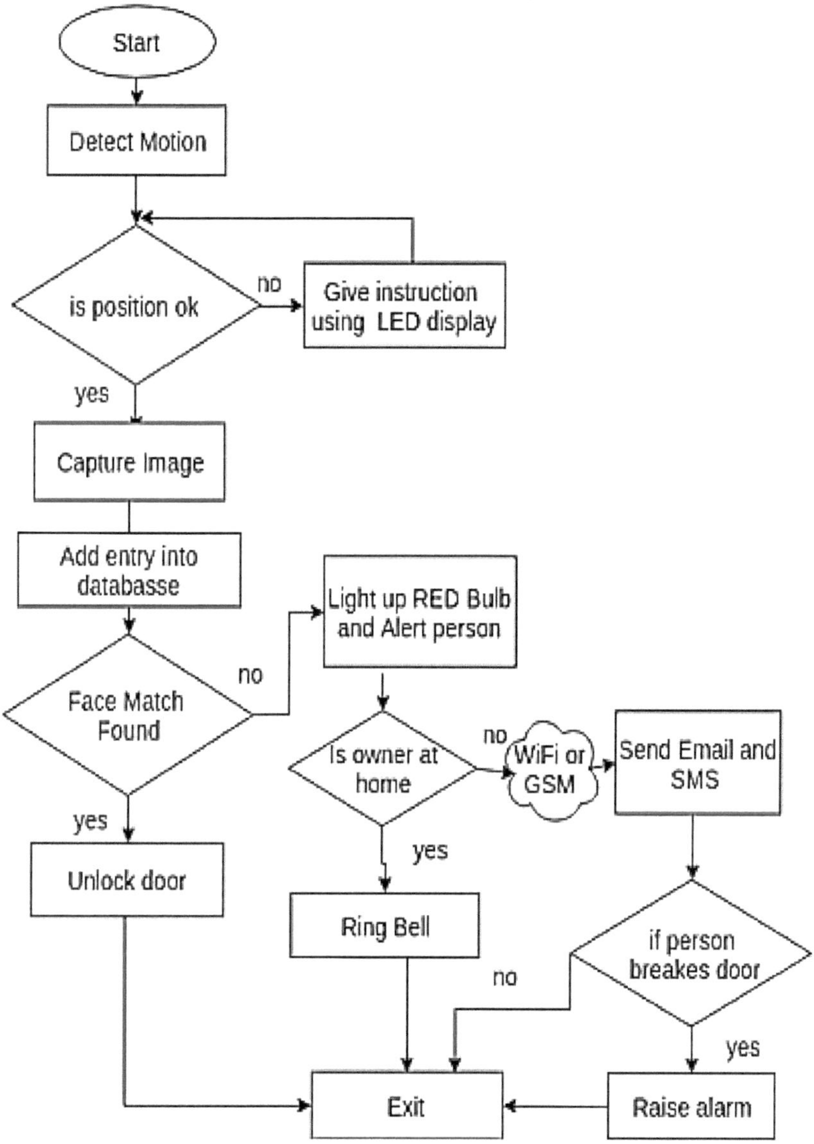

FIGURE 9.4 System flow diagram for face detection.

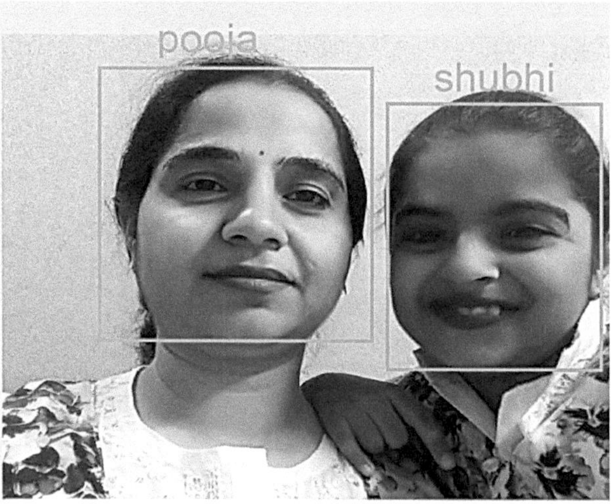

FIGURE 9.5 Face recognition results for registered (known) faces.

Figure 9.6 shows that show the result of the face recognition process such that the system recognized determined it as a known person and it also shows his/her name on an image by covering a rectangular box on the face.

FIGURE 9.6 Face recognition results for unknown (not registered) faces.

9.6.2 MESSAGE GENERATION

Figure 9.7 shows a message when someone keeps a photo of a person in front of the home/office camera then an automatically generated message is sent to u r mobile no from the system.

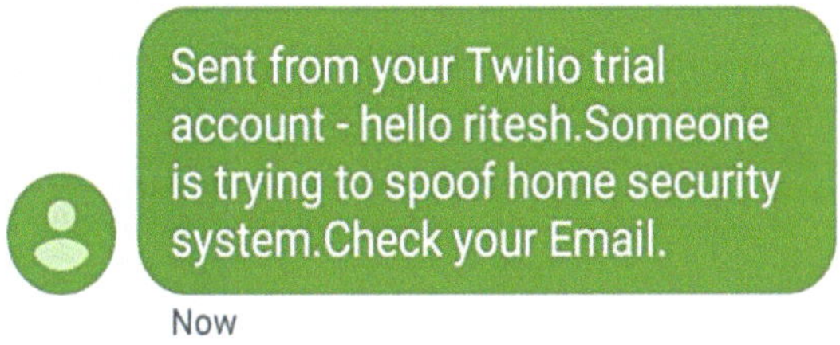

FIGURE 9.7 The message generated for spoofing.

Figure 9.8 shows a message when some unknown person comes in front of the home/office camera then an automatically generated message sent to u r mobile no from the system.

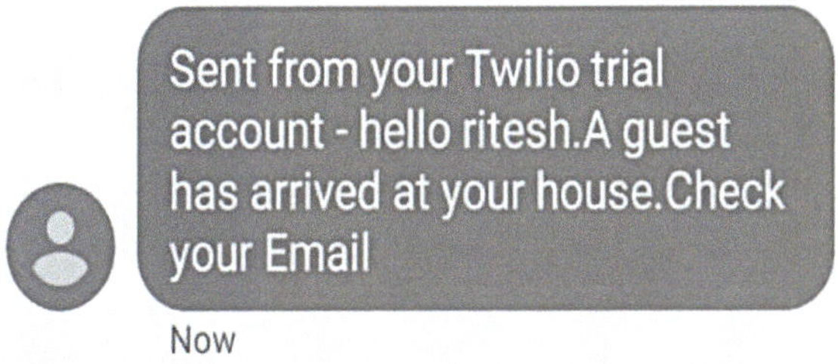

FIGURE 9.8 Message generated for new guest.

Figure 9.9 shows a message when LPG gas leakage occurs, then an automatically generated message sent to u r mobile no from the system.

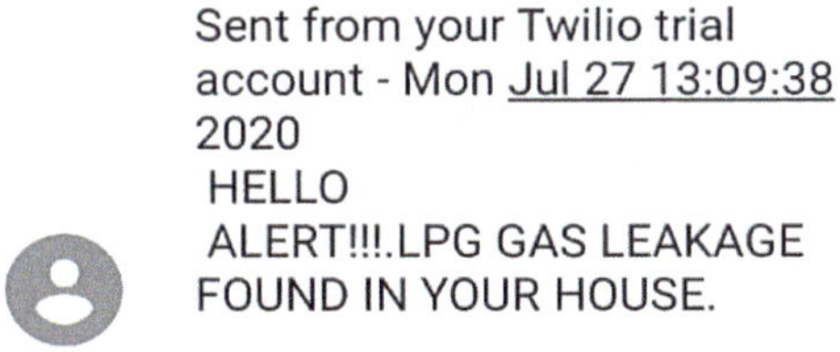

FIGURE 9.9 Message generated for LPG leakage.

Figure 9.10 shows a message when smoking activity is detected by the system then an automatically generated message sent to u r mobile no from the system.

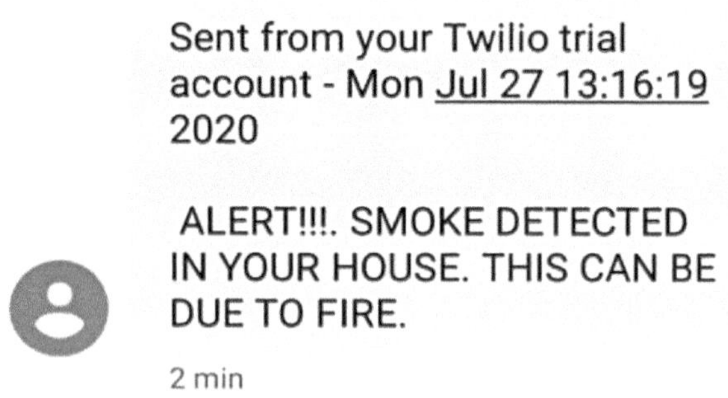

FIGURE 9.10 Message generated for smoke detection.

Figure 9.11 shows a message when some electronic appliances are left on in the office or home place, then an automatically generated message sent to u r mobile no from the system.

FIGURE 9.11 Message generated for appliances on/off status.

9.6.3 MAIL GENERATION

Figure 9.12 shows mail detail when someone keeps a photo of a person in front of home/office camera then an automatic generated e-mail sent to the e-mail id from the system.

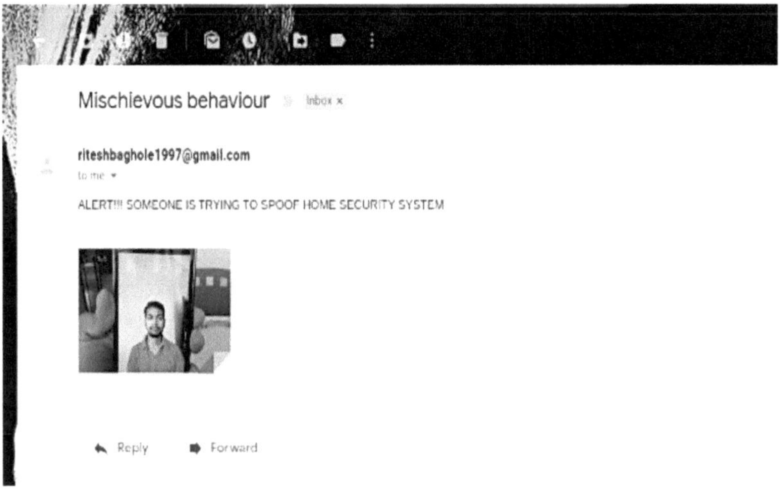

FIGURE 9.12 Mail generated for spoofing.

Figure 9.13 shows mail detail when some unknown person comes in front of the home/office camera, and then an automatically generated mail is sent to your e-mail id system.

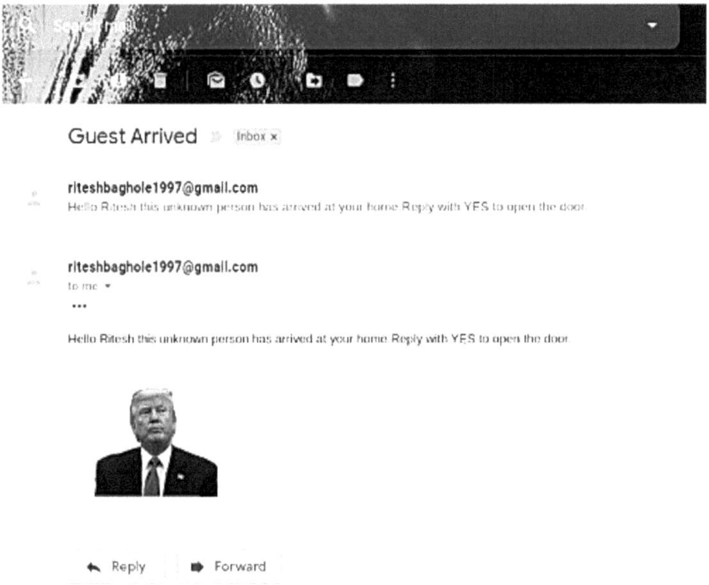

FIGURE 9.13 Mail generated message for an unknown person.

9.7 CONCLUSION

In this work, an intelligent security system is implemented through the IoT. It is a powerful way to open a door/office door if the person at the door who is the authorized member can zoom in on camera with detailed viewing for health checks and separate secure feature support like LPG leak and smoke detector. We can check local electronic status anywhere. The manager can use it anywhere to reduce electrical wastage. This system requires minimum user interaction and is user-friendly. It can be easily installed and can be used in houses, banks, etc., where high security is needed. The system operates efficiently and is easily controllable. The work has produced satisfying results.

KEYWORDS

- **architectural design**
- **face recognition**
- **intelligent security system**
- **Internet of Things (IoT)**
- **python languages**
- **Raspberry Pi**
- **sensor**

REFERENCES

1. Bhatia, P., Rajput, S., Pathak, S., & Prasad, S., (2018). IoT based facial recognition system for home security using LBPH algorithm. In: *2018 3 International Conference on Inventive Computation Technologies (ICICT)* (pp. 191–193). IEEE.
2. Gupta, P., & Chhabra, J., (2016). IoT based smart home design using power and security management. In: *2016 International Conference on Innovation and Challenges in Cyber Security (ICICCS-INBUSH)* (pp. 6–10). IEEE.
3. Gupta, A. K., & Johari, R., (2019). IoT based electrical device surveillance and control system. In: *2019 4 International Conference on Internet of Things: Smart Innovation and Usages (IoT-SIU)* (pp. 1–5). IEEE.
4. Vadivukarasi, K., & Krithiga, S., (2018). Home security system using IoT. *International Journal of Pure and Applied Mathematics, 119*(15), 1863–1868.

5. Adiono, T., Fuada, S., Anindya, S. F., Purwanda, I. G., & Fathany, M. Y., (2019). IoT-enabled door lock system. *Int. Journal of Advanced Computer Science and Applications, 10*(5), 445–449.
6. Malche, T., & Maheshwary, P., (2017). Internet of things (IoT) for building a smart home system. In: *2017 International Conference on I-SMAC (IoT in Social, Mobile, Analytics, and Cloud) (I-SMAC)* (pp. 65–70). IEEE.
7. Chifor, B. C., Bica, I., Patriciu, V. V., & Pop, F., (2018). A security authorization scheme for smart home internet of things devices. *Future Generation Computer Systems*, *86*, 740–749.
8. Mohan, M., Kannan, V., & Keshav, M., (2017). Phone controlled security lock systems. *Int. J. Adv. Res. Comput. Commun. Eng. (IJARCCE), 6*(4), 155–160.
9. Kodali, R. K., Jain, V., Bose, S., & Boppana, L., (2016). IoT based smart security and home automation system. In: *2016 International Conference on Computing, Communication, and Automation (ICCCA)* (pp. 1286–1289). IEEE.
10. Patchava, V., Kandala, H. B., & Babu, P. R., (2015). A smart home automation technique with raspberry pi using IoT. In: *2015 International Conference on Smart Sensors and Systems (IC-SSS)* (pp. 1–4). IEEE.

CHAPTER 10

Intelligent Agriculture System

ASHWINI A. WAOO and AKHILESH A. WAOO

Department of Biotechnology, FLST, AKS University, Satna, Madhya Pradesh, India

Department of CS/IT, FE&T, AKS University, Satna, Madhya Pradesh, India

ABSTRACT

Integrated and interdisciplinary technology is now the prime wealth of knowledge in all aspects of life. Interdisciplinary research brings new opportunities for the benefits of short-term projects. Agriculture is the main foundation of the Indian economy; it is possible to sustain the agricultural system with innovative interdisciplinary approaches so that India will emerge as a "Superpower" of the world. The present chapter proposal is a prototype demonstrating agriculture engineering at a mass scale with an intelligent system and minimum expenditure.

Various factors, such as drastic population growth, climate change, global warming, and food security concerns, have driven the research communities to seek more innovative approaches for improving crop yield and food protection. This content will explore applications of artificial intelligence (AI) for the agricultural sector and its modernization.

10.1 INTRODUCTION – IAS

The transformation of traditional agriculture into smart agriculture makes it more efficient. The discovery of robots and drawings helps farmers with large

Intelligent Sensor Node-Based Systems: Applications in Engineering and Science, Anamika Ahirwar, Piyush Kumar Shukla, Prashant Kumar Shukla, and Ruby Bhatt (Eds.)

production in less time. This chapter deals with issues of traditional agriculture and the problems associated with traditional agricultural management and how these problems were found by the application of new technology in agriculture with the introduction of an intelligent system for agriculture, that is, IAS. This chapter focused on future designing tools of IAS (Figure 10.1), such as agricultural robots for performing agricultural tasks, IAS-based micro-irrigation control technology, monitoring of high voltage electrical devices used in agriculture, continuous crop and soil monitoring for better agricultural yield, IAS-based weather monitoring, predictive data analytics for IAS. The last section of this chapter discusses the aspects of the Indian perspective regarding IAS implementation and monitoring.

FIGURE 10.1 Tools of intelligent agriculture system.

Source: Adapted from https://www.clipartmax.com/middle/m2H7G6G6N4H7G6A0_tractor-farm-field-agriculture%02planter-farm-field-farm-clipart/

10.2 TRADITIONAL AGRICULTURE

Indigenous farmers and rural peoples of India follow traditional/conventional agricultural techniques such as intercropping, cover cropping, agroforestry, crop rotation, and composting. All these traditional practices enhance crop productivity and also help to minimize the climate change challenges and overall crop yield.

Traditional knowledge is holistic because it has multiple applications in various fields, such as climate, soil science, hydrology, agriculture, human health, and animal welfare. Traditional agriculture is mainly based on the experiences of previous agricultural practices performed by local farmers for thousands of years. It plays a significant role in building scientific knowledge of agriculture.

Several pieces of research are carried out all over the world in agriculture because the total available irrigated land is rapidly reduced in the world, and there is a continuous and exponential increase in demand for food resources. Global warming is also a new major challenge for good and standard agricultural practices [1]. Various problems are faced by traditional agricultural practices, and this section describes these problems in detail.

10.2.1 SOIL SALINIZATION

Continuous irrigation and over-irrigation of agricultural fields lead to soil salinization problems. This problem is widespread in various regions of India as well as in the world. Soil salinization is caused due to the accumulation of a high concentration of salts in the soil, which increases the salinity of the soil. It results in reducing the crop yield and gradually makes that soil infertile in the future.

10.2.2 LOSS OF AGRICULTURAL LAND DUE TO URBANIZATION

Due to industrialization and urbanization, various agricultural fields are converted into urban areas and industries. It leads to the reduction in total agricultural land that causes the problem of loss of agricultural land for farming. It is very difficult to harvest large amounts of crop yield from small agricultural land. To fulfill the demand of the increasing population, this is also one of the major challenges of traditional agriculture in the future.

10.2.3 WATER DEPLETION

The increasing population worldwide poses the problem of water depletion in various areas of the world; traditional agricultural practices require more amount of water due to reduced rainfall and global warming groundwater levels being minimized.

10.2.4 POLLUTION AND SILT

Rapid industrialization and urbanization cause lots of soil pollution as well as air pollution. Water carries soil pollution to the water, and there is a problem of accumulation of pollution and slate in the water; such polluted water is not suitable for any type of plantation, and heavy sleet load causes harmful effects to the plants as well as animal life.

10.2.5 PESTICIDE RESISTANCE

To fulfill the increasing demand of the population, there is a green revolution, but with the onset of the Green Revolution, the major problem arises from the extensive use of herbicides, pesticides, and insecticides. Due to the heavy load of these chemicals, most of the paste and herbs develop a natural resistance towards these chemicals, and that is a problem for agricultural practices. Even these types of chemicals result in barren soil, ultimately decreasing the agricultural land.

10.2.6 FERTILIZERS AND EUTROPHICATION

The green revolution in traditional agriculture brings extensive use of chemical fertilizers. Most of these chemical fertilizers enter into the hydrological cycle to water runoff. Does water become and reach weight dissolved nutrients such as phosphates and nitrates? Due to increased levels of these nutrients, the growth of aquatic plants in the water resources is overstimulated, and due to enormous growth, there is a depletion of dissolved oxygen levels inside the water body. This overall process is called eutrophication. Eutrophication harms aquatic life and the aquatic ecosystem.

10.2.7 EROSION

A major amount of soil loss has been caused due to erosion all over the world from ancient times. If the soil is eroded, it becomes productive. The natural vegetative cover on soil protects it from the Eros agents such as wind, water, ice, or gravity. Also, the roots of trees hold the rainwater and prevent soil erosion. In traditional agriculture, soil erosion is a common problem faced by farmers.

10.3 INTELLIGENT SYSTEM FOR AGRICULTURE (IAS)

Intelligent system for agriculture (IAS) is the re-orientation of traditional farming to increase yields through enhanced productivity through automated technology. Here are some advantages of an intelligent system for agriculture:

- Economic farming;
- Sustainable use of natural resources;
- Highest productivity;
- Energy conservation;
- Environmental protection;
- High income and employment generation;
- Speed-up of the farming process;
- Improved accuracy;
- Overcome climatic challenges.

The intelligent agricultural sector works for various parameters in agricultural field management, mineral mapping, soil testing mapping, and weather forecasting. These can be measured by using some software solutions involving artificial intelligence (AI), data analytics, and machine learning. Intelligent Agricultural systems are based on smart technologies reducing the side effects of traditional farming.

10.4 FUTURE DESIGNING TOOLS OF IAS

Automation technology will change farming organizations and management practices. Automated tractors and electric equipment and instruments will enable farmers to work simultaneously over a large area of an agriculture field in a day and night continuous manner. Automated irrigation systems help to collect information about soil and water specifications which convert the agronomic activities in a very efficient manner.

A simple smartphone can be aided with some specialized agricultural applications to monitor the continuously changing climate along with soil and water parameters in agriculture, as shown in Figure 10.2.

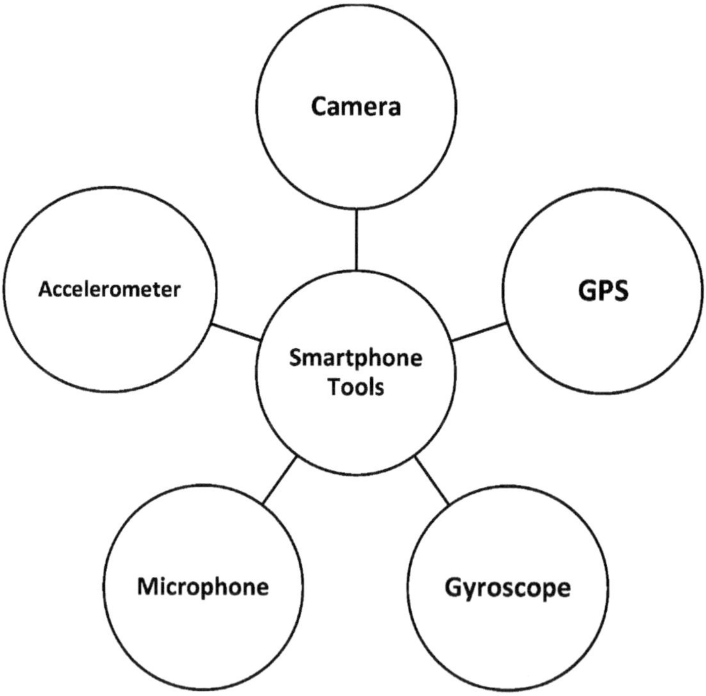

FIGURE 10.2 Smartphone tools for agricultural information and data analysis.

Various smartphone applications with the internet of things (IoT), and data aggregation, are now available for farmers, which can provide rapid response and processes to carry out agronomical activities like seeding, fertilizing, weeding, and irrigation. These software systems record and collect data from weather stations and small remote sensing devices. These all work very well for larger agricultural land, but now several apps have been designed to target small-scale farmers. The following are some issues that are targeted to make apps for farmers:

1. **Soil Study:** Soil images are used to monitor and adjust changing soil parameters like temperature, pH, etc.

2. **Disease Diagnostics:** Images of suspicious plants can be forwarded to professionals for diagnosis.
3. **Water Requirement Analysis:** Leaf characteristics from photos and brightness logging can determine water requirements.
4. **Fertilizer Calculator:** Soil sensors and leaf color measure the requirements for nutrients.
5. **Crop Harvesting:** Photos with white lights and UV precisely forecast the ripening period of crops over a large agricultural field.

A detailed discussion on various new technology-based tools in smart agriculture is given in subsections such as – Agribots, drones, monitoring of high voltage electrical devices, smart micro-irrigation technology, etc.

10.4.1 AGRICULTURAL ROBOTS

Fully automated robots are very important. They play important activities and can help human labor management. This evolution of the role of robots potentially changes the new job roles of human beings, which are more technical. Agricultural robots are mainly called Agribots, these are based on AI, and they have promising prospects for the agricultural area with enhanced execution across the larger segment of agriculture and agricultural-based industries.

Digitized data robots accurately identify right versus Android plants in the greenhouse and can establish precise harvesting systems for them and also collect them in an exact on the World Boxing system in such a way harvesting and Packaging of fruits and greenhouse products can be made easy for farmers.

Intelligent system-based agriculture is transforming into a high-tech industry and creating opportunities for new professionals, new companies, and new investors. This technology rapidly developed and formed new avenues for farmers. The major role of the transformation of agriculture into a high-tech industry is advanced robotics and automation technology.

The application of Agribots increased the production yield for farmers. It is necessary for future agriculture to search for innovative applications and automatic Technologies. Agribots replace monotonous and slow tasks for farmers, and this can allow more targeted work toward the increased overall production yield.

Changing the traditional farming practices in crop harvesting is a significant agricultural task that can be automated with the help of an agricultural robot. When consumer demand is higher and labor requirement is greatest, then automation will improve the fulfillment of demand. This automation in technologies can improve sensitivity and operation capacity in the agricultural field. Agribots and Drone's management also works to maintain safety considerations and the full-proof effectiveness of the working drones.

In the agricultural sector, drones are used in place of the human liver. These loans are equipped with multispectral photo cameras, which can easily monitor the parameters like plant growth and crop stress, and they can also help in the prediction of Agricultural aid by generating a vast amount of digital data. Some more advanced drones are also able to carry herbicides, water, and fertilizer at the specific site in the agricultural area. The drones can be applied via a drone service operation with scheduled commands or can be gathered on-site and utilized as weatherproof docking stations to maintain the drones for recharge and send data for analysis (Figure 10.3).

FIGURE 10.3 Drones in agricultural field.

Source: Adapted from https://publicdomainvectors.org/en/free-clipart/Dronepilot/45288.html

Management of drones requires a weatherproof docking station that can manage drones by recharging them and storing the data from drones for analysis here, and drones can be scheduled for a particular agricultural task. Agribots and Drones perform various tasks in agriculture. Some of the functions are listed in subsections.

10.4.1.1 SPRAYING AND WEEDING

Drones can generate databases like weeds measurement and images in the agricultural area, and then these databases can be used for agricultural sites for applying pesticides and weedicides on agricultural land. In this way application of pesticides is directly on the weeds, and it does not harm the agricultural crop.

10.4.1.2 AUTO STEERING

GPS enables farmers to run their tractors at a specific write-in agricultural area in coordination with other farm equipment, and this results in an operations task that can be done without any fatigue farmers can operate this type of auto-steering vehicle from a remote area with no need to drive.

10.4.1.3 FRUIT AND GREENHOUSE HARVESTING

Specially designed robotic arms can help farmers in harvesting the crop, especially fruits and harvested by this harvesting robotics. Such type of harvesting robots is primarily used in the fields of strawberry Orchid fruit like apples and cucumbers.

10.4.2 IAS-BASED MICRO-IRRIGATION

The proper irrigation system is required for the maintenance of crops. The micro-irrigation system based on soil moisture sensor value can be developed nowadays by using a smart agricultural Technologies irrigation system that uses wild and sprinklers, which are connected to the controller using a relay switch. Using search micro-irrigation techniques, precise amounts of water can be supplied at the roots of crops according to the need of individual agriculture in such a way water conservation can be possible, and plants get the optimum moisture for their growth [2].

Traditional irrigation methods are generally not suitable for water conservation, but modern technological advances can be made to reduce the wastage of water. Smart agriculture and micro-irrigation technology have some advantages over the traditional irrigation strategy. They are as follows:

- Reduce water consumption and wastage;
- No human labor required;
- Reduces soil erosion;
- Reduces nutrient leaching;
- Cost-effective method for long-term use;
- Damage proof system against weather and birds;
- Increased productivity of farmland.

Micro-irrigation technology is based on sensors that sense the parameters like temperature and moisture from a specific location, and this data is collected by sensors used for monitoring the crops. By using various algorithms, threshold values can be set up for this micro-irrigation system, and according to a change in the threshold value, micro-irrigation Technology performs precise and effective water management in agricultural land. The micro-irrigation system can be powered and scheduled through a web page. In the micro-irrigation system sensor network was developed precisely. In a complex environment, data is collected every single time, and the micro-irrigation monitoring system works for information collection and monitoring. This system manages the water requirement of the agricultural field efficiently.

The system can be powered by having a communication link on a cellular interface that allows monitoring of data and irrigation scheduling programs through a web page. The advanced system with novel technologies in agriculture provides benefits for farmers in the form of an increase in agricultural yield. A remote sensing system to control irrigation using a distributed wireless sensor network (WSN) was implemented for the same.

10.4.3 HIGH VOLTAGE ELECTRICAL EQUIPMENT

High-voltage equipment is a crucial part of the power system, like transformers, circuit breakers, and switchgear. The smooth operation of these devices required a guaranteed proper supply of electricity. The monitoring and controlling of high-voltage equipment is an important part of routine maintenance in smart agriculture. The monitoring of high-voltage equipment based on IoT using remote sensors and wireless networks has a thrust area of research. The major advantage of high-voltage equipment is that it saves manpower. It is necessary to develop a remote monitoring system for regulating leakage currents as well as ground currents in high-voltage electrical substations. Wireless local area networks with low power consumption and overvoltage protection [3] can be implemented to monitor a variety of crops within the substation.

10.4.4 CROP AND SOIL MONITORING

Continuous monitoring is essential for better agricultural yield. In a Smart agricultural system based on various sensors that detect minor changes, Airflow Sensors detect soil air porousness. Sensor senses in moving conditions in particular areas. Various soil properties, such as soil type, structure, compaction, and dampness status, result in interesting remarks.

Electrochemical sensors are used to sense accuracy in horticulture. These sensors measure pH and soil nutrient levels. Sensors mounted in a specific area according to need. Mechanical Sensors are used to detect soil compaction. These sensors utilize the detection of records. Sensors of soil moisture measure dampness status by detecting the dielectric steady in the dirt. Wind current Sensor nodes work for soil-air porousness measurement. Yield Monitoring frameworks can be set on crop-collecting vehicles. They harvest yield by time, in particular, GPS area estimated and recorded. Fertilizer application instruments are used to yield optical studies of the plant. The following are some arrangements in the field using smart technological equipment [4]:

1. **Variable Spraying:** These controllers were used for herbicide spray after identifying weed locations from the map. The volume and mixing of the spray can be automatically determined.
2. **Weed Mapping:** It is used to construct maps by identifying the location with a special GPS receiver. Maps can be used multidimensionally for yield, spray, and fertilizer status.
3. **Topography and Boundaries:** It can be recorded easily with GPS. These precision maps are useful for the interpretation of yield and weed maps. Field areas, wetlands, and roads can be precisely mapped in farm area planning.

4. **Guidance Systems:** It can precisely mark the position of a moving vehicle around 30 cm using GPS. It replaces traditional vehicle operations for seeding or spraying.
5. **Salinity Mapping:** It can be done with a salinity meter which interprets the change in salinity over time.

10.4.5 IAS-BASED WEATHER MONITORING

Weather station plays a significant role in the cultivation and monitoring of economically important crops like tea, coffee, grapes, potatoes, etc. Individual farmers and farming corporations at present applying automatic weather stations for the following advantages:

- Weather sensors allow us to identify an ideal region to grow a particular cash crop;
- Protecting crops in contradiction of adverse climatic conditions;
- It identifies plant diseases and manages pest and insect intrusion.

Information received from the weather station is used for the study to enhance irrigation patterns and pest management using IAS based weather monitoring system (Figure 10.4).

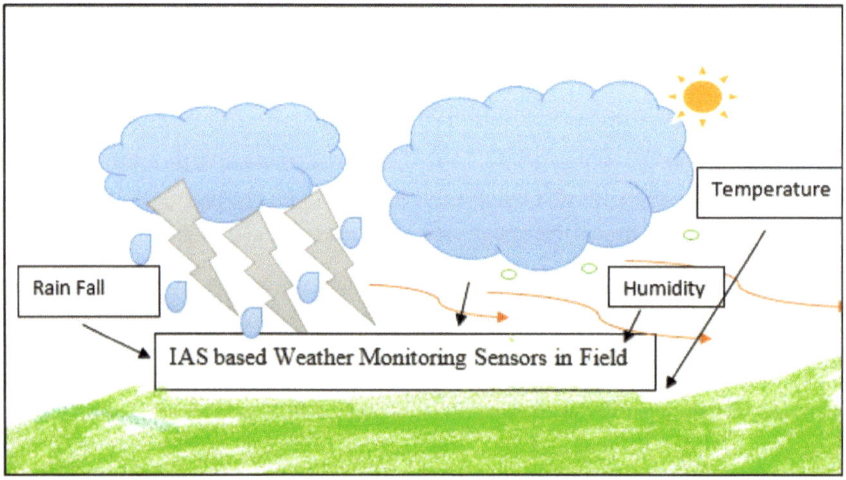

FIGURE 10.4 IAS-based weather monitoring.

Agricultural insurance in India is a huge market, and insurance companies will rely on the data received from weather stations to resolve or refuse the farmer's claim [5]. The following are some critical weather monitoring parameters:

1. **Wind:** Its direction and speed are important parameters to warn farmers of a coming storm and other disasters.
2. **Temperature:** Detecting changes in temperature during a specific period gives predictive conditions for crops.
3. **Humidity:** Its measurements help in preparing for rain so they can use water smartly.
4. **Rainfall:** It can be predicted by analyzing all historical data over defined periods which gives worthful input for predictions using AI-based algorithms.
5. **Air Pressure:** It is an important factor for predicting weather changes.

10.4.6 PREDICTIVE DATA ANALYTICS

The actual use of digital records related to agricultural practices reveals a highly efficient aspect of the documentation of the various types of strategies, which can be implemented for finding the aspects of a given sale [6]. Thus, multinational companies are important for fresh, budding trade crops and their integration into the global supply chain of crops.

10.5 IAS IMPLEMENTATION – INDIAN PERSPECTIVE

Agriculture has a very crucial role in the economy of any country and especially in the developing country of India. Smart agriculture intelligent agricultural system exploits automation in the field of agriculture to sustain the economy as the population growth is exponential across the country. The food and employment demand is also increasing, and the traditional method there are not sufficient enough to fulfill all these demands. Automation and newer Technologies must be used by farmers to increase production, and ultimately it results in a strong economy for the nation. New approaches to intelligent agriculture systems (IAS) sustain the food requirements and also provide new employment opportunities in the field of agriculture as well as agricultural industries.

There are many challenges and advantages of incorporating smart agricultural students in India. Farmers are taking an interest in these new technologies. It can be said that India is going into the future of agriculture. Smart technology in agriculture enables farmers to Increase profit, reduce waste and sustain better environmental quality during and after agricultural practices. New farming technology requires professional skills. The farmer is not a person with a passion for agriculture, and he is a trained person with professional skills to handle smart tools based on IoT [7], machine learning, and AI. This new perception of farmers and farming will be transformed into new areas for successful entrepreneurs. Agriculture act as a pillar of the Indian economy in which villages are considered the lifeline of India [8]. Around 70% of rural and 8% of urban households are dependent on agriculture for employment.

India is developing precision agriculture and smart agriculture far more slowly than European nations. Indian farmers have very small farms, and the expense of installing automatic equipment is highly expensive and impractical [9]. The Indian government has started a number of initiatives employing artificial intelligence (AI) and the internet of things (IoT) in agriculture. Funding is provided to state governments for digital agriculture projects using emerging technologies like artificial intelligence and machine learning (AI/ML), the internet of things (IoT), blockchain, etc. This funding is available through the National e-Governance Plan in Agriculture (NeGPA) program with the Department of Agriculture and Farmers' Welfare. But still, more time will be needed to bring this technology to individual Indian farmers [10].

10.6 ADVANTAGES OF IAS IN AGRICULTURE

An intelligent system embedded in agriculture helps farmers in various ways, such as weather forecasting, water management, automation in agronomical practices, and enhancing crop yield. IAS does not eliminate the jobs of human farmers, but it will improve their difficult processes and enhance production. The following are the main advantages of IAS [11]:

- Efficient processes for the production, harvesting, and selling of cash crops.

- Implementation of AI to check diseased and defective crops and increase the efficiency of healthy crops.
- Efficient maintenance and monitoring of agro-based industries and businesses with AI technology.
- Automated machine adjustments, robots, and drones for different aspects of agricultural practices.
- Improvement in crop management practices by AI [12].

10.7 SUMMARY

The digital era of agriculture has the potential to establish sustainable agricultural practices with the least investment and smooth and accurate monitoring that proved to be a precious gift for mankind. AI or machine-learning technologies improve crop yield and face challenges such as soil health, disease occurrence, and global warming [13]. Digital agriculture is the way of smart farming and is the basis of intelligent decision systems for agriculture engineering. This transformation of traditional agriculture to digital agriculture can create the possibilities of water management, pollution-free agricultural practices, and stable crop yield despite adverse climatic challenges [14]. IAS-based farming systems accept the challenges of natural parameters and stabilize them by using embedded technology for the entire management of agricultural production to fulfill the demand of the nation.

KEYWORDS

- **agriculture**
- **artificial intelligence**
- **crop yield**
- **digital agriculture**
- **interdisciplinary research**
- **Internet of Things**
- **machine learning**

REFERENCES

1. Ali, I., Jianmin, G., Naz, T., Chandio, F., Ali, N., & Qureshi, W., (2018). Monitoring and control systems in agriculture using intelligent sensor techniques: A review of the aeroponic system. *Sensors and Embedded Systems in Agriculture and Food Analysis*, doi.org/10.1155/2018/8672769.
2. Ramachandran, V., Ramar, R., & Srinivasan, S., (2018). An automated irrigation system for smart agriculture using the internet of things. *Conference: 2018 15 International Conference on Control, Automation, Robotics and Vision (ICARCV)*. doi: 10.1109/ICARCV.2018.8581221.
3. Chen, J., Tao, J., & Huo, L., (2018). Study on communication methods for electric power high-voltage equipment monitoring system. *International Journal of Online Engineering, 14*(02), 181. doi: 10.3991/ijoe.v14i02.8212.
4. Schriber, S., (2019). *Smart Agriculture Sensors: Helping Small Farmers and Positively Impacting Global Issues, Sensor Technology*. https://www.techbriefs.com/component/content /article/tb/supplements/st/features/applications/34585 (accessed on 21 December 2022).
5. Tenzin, S., Siyang, S., Pobkrut, T., & Kerdcharoen, T., (2017). Low-cost weather station for climate-smart agriculture. In: *9 International Conference on Knowledge and Smart Technology (KST)* (pp. 172–177). Chonburi. doi: 10.1109/KST.2017.7886085.
6. Yasam, S., & Nair, A., (2019). Precision farming and predictive analytics in agriculture context Srinath. *International Journal of Engineering and Advanced Technology (IJEAT)* (Vol. 9, No. 1S5, pp. 74–80). ISSN: 2249 – 8958, Blue Eyes Intelligence Engineering & Sciences Publication. doi: 10.35940/ijeat.A1023.1291S52019.
7. Gu, W., & Yuan, H., (2016). Research on IoT technology applied to intelligent agriculture. In: Juntao L., Huang, B., & Yao, Y., (eds.), *Proceedings of the 5 International Conference on Electrical Engineering and Automatic Control* (pp. 1217–1224).
8. Faggella, D., (2020). *AI in Agriculture – Present Applications and Impact*. Retrieved from https://emerj.com/ai-sector-overviews/ai-agriculture-present-applications-impact/ (accessed on 21 December 2022).
9. Tamilnadu, R., & Vinith, P., (2015). Intelligent agricultural system with weather monitoring. *IJSRD - International Journal for Scientific Research & Development, 3*(10). ISSN (online): 2321-0613.
10. Singh, R., & Singh, G. S., (2017). Traditional agriculture: A climate-smart approach for sustainable food production. *Energ. Ecol. Environ., 2*(5), 296–316. doi: 10.1007/s40974-017-0074-7.
11. Gupta, J., (2019). *The Role of Artificial Intelligence in Agriculture Sector.* Blog, Enterprise Technology. Retrieved From https://customerthink.com/the-role-of-artificial-intelligence-in-agriculture-sector/ (accessed on 21 December 2022).
12. Sennaar, K., (2019). *Agricultural Robots – Present and Future Applications*. https://emerj.com/ai-sector-overviews/agricultural-robots-present-future-applications/ (accessed on 21 December 2022).

13. Amalraj, J., Banumathi, S., & John, J., (2019). A study on smart irrigation systems for agriculture using IoT. *International Journal of Scientific & Technology Research, 8*(12). ISSN: 2277-8616 1935 IJSTR©2019 www.ijstr.org.
14. Kamble, D. B., (2009). *Precision farming in Indian Agricultural Scenario, Remote Sensing and Geographic Information System*. Geospatial World.

CHAPTER 11

Intelligent and Smart Agriculture System Using Cooperative Approach

BHUPESH GOUR and JAY PRAKASH MAURYA

Department of Computer Science and Engineering, Lakshmi Narain College of Technology, Bhopal, Madhya Pradesh, India

ABSTRACT

Fertility of the soil, weather, crop growth, temperature, rainfall, and information on seed planting, among other things, are significant parameters to consider for the development and improvement of Indian agriculture. All parameters can be gathered via IoT sensors and digital devices and stored in real-time database environments for sharing with digital machines. It aids farmers in obtaining information on all aspects of agriculture. Agriculture processes may be monitored using sensors, smart cameras, mobile applications, and gadgets such as microchips, thanks to internet technology. The automated technology provided by the internet of things (IoT) assists farmers in a variety of ways, including the most efficient use of resources (resources are finite) and agricultural problems.

11.1 INTRODUCTION

The global population has quadrupled in the last century. There were 1.8 billion people on the planet in 1915. According to the most recent UN estimate, there are 7.3 billion people on the planet today, with a potential of 9.7 billion by 2050. Global food demand is rising as a result of this

Intelligent Sensor Node-Based Systems: Applications in Engineering and Science, Anamika Ahirwar, Piyush Kumar Shukla, Prashant Kumar Shukla, and Ruby Bhatt (Eds.)

growth, as well as improving financial conditions in developing countries. Nothing can avoid a significant increase in the global death rate at this late stage. Furthermore, many skeptics believe Ehrlich's claim is exaggerated, claiming that human population growth follows an exponential pattern. However, according to the principles of nature, exponential growth cannot be approached indefinitely (Figure 11.1).

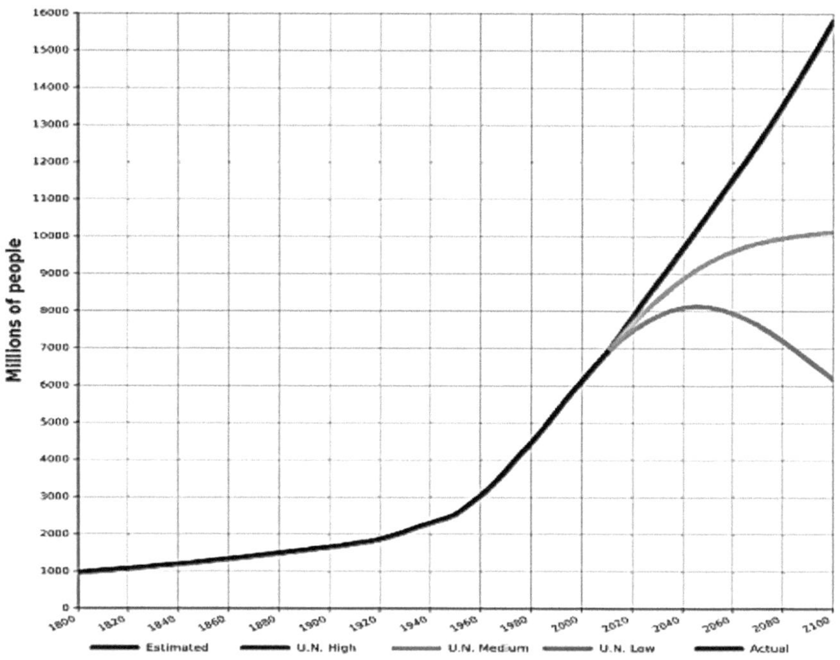

FIGURE 11.1 World population growth from 1800 to 2100.

Source: Wikimedia Commons. https://creativecommons.org/licenses/by-sa/3.0/deed.en

Aside from food shortages, population growth is doing havoc on the ecosystem in a variety of ways that are irreversible. Many scientists agree that global climate change is caused by carbon dioxide (CO_2) emissions and is a substantial result of human activity. Throughout the late twentieth century, a succession of accords was signed for commitment to reducing their CO_2 emissions to stop the rise in global warming; however, not every government has ratified these treaties, owing to economic and political considerations. In some circles, the role of the act in global climate change is strongly contested. For limiting human population growth and conserving the ecosystem in the long run, there is a lot of uncertainty.

Food demand is predicted to extend from 59% to 98% by 2050. This will necessitate agricultural market structure in ways we haven't seen before. Farmers must participate in raising crop production, either by expanding agricultural land to grow more crops or by enhancing productivity in existing agricultural areas through fertilizer, irrigation, and the use of new technologies or techniques such as precision farming. The expected annual growth rates of global demand for important agricultural commodities are depicted in Figure 11.2. Meat, fish, roots, and tubers demand is expanding at around half the rate of the previous decade and not quite as fast as the global population growth rate of around 1% per year. The decline is much more pronounced for vegetable oils, which had experienced rapid expansion in the previous decade. The exceptions are fresh dairy, which is seeing an increase in demand, and sugar, which is expected to expand at a similar rate.

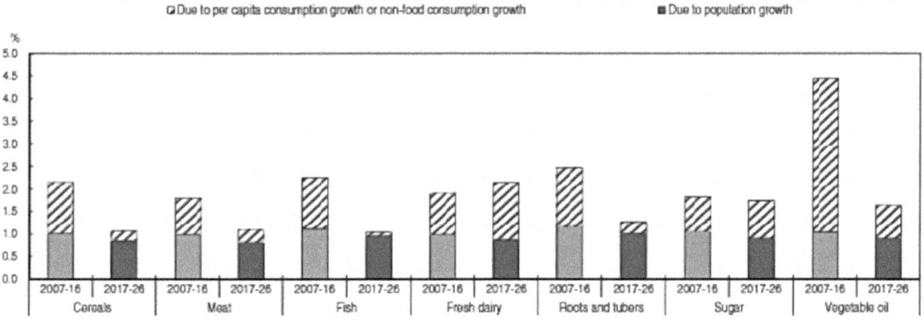

FIGURE 11.2 Annual growth rates of demand for key agricultural commodities, 2007–2016 and 2017–2026 [2].

Source: Adapted from https://voxeu.org/article/demand-agricultural-commodities-grow-more-slowly-next-decade

However, particularly in tropical regions, the ecological and socioeconomic costs of clearing more land for agriculture are typically considerable. Crop yield, or the amount of crops harvested per unit of cultivated area, is now increasing steadily in order to satisfy projected food demand (Figure 11.3).

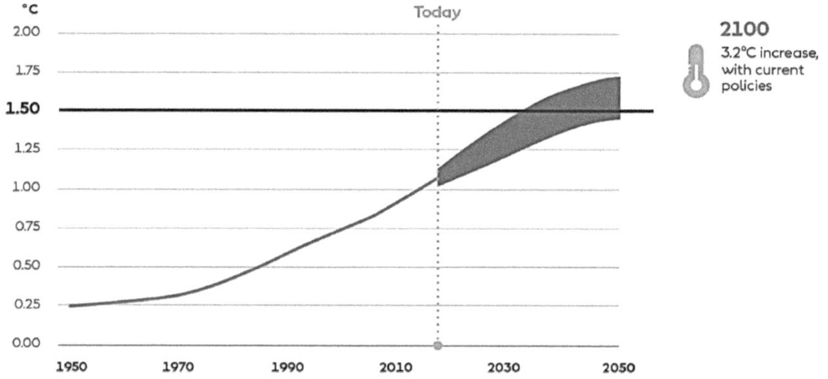

FIGURE 11.3 Temperature change growth.

More issues, such as global climate change, urbanization, and a lack of investment, will make meeting food demand difficult. Climate change-related water scarcity, rising global temperatures, and extreme weather will have serious long-term effects on crop output, according to an academic agreement. Many major agricultural regions, particularly those near the Equator, are affected. Due to global climate change, the Brazilian state of Mato Grosso, which is one of the world's most important agricultural regions, will see an 18% to 23% decline in soy and corn output by 2050.

Extreme heat can also cause significant reductions in agricultural productivity in the Midwest of the United States and Eastern Australia, two additional globally important regions. Despite the fact that certain places will initially benefit from global climate change, northern countries, such as China, Canada, and Russia, are expected to have longer and warmer seasons in certain areas.

Due to substantial crop yield gaps and broad abandoned farmland (more than 40 million hectares, an area greater than Germany) following the fall of the Soviet Union in 1991, Russia is already a significant grain exporter with a huge untapped output potential. This country offers the best agricultural opportunity on the planet, but to take advantage of it, institutional reform and large investments in agriculture and rural infrastructure are required. Advanced logistics, transportation, storage, and processing are also required to ensure that food is transported from places where it grows abundantly to places where it does not. Soft commodity trading organizations, such as Cargill, Louis Dreyfus, or COFCO, are frequently in demand.

While large food corporations such as General Mills and Unilever have a global impact on what people eat, trading companies have a direct impact on food security because they are the source and distributor of staple foods and ingredients used by Big Food, such as rice, wheat, corn, and sugar, as well as soybeans. They also store grains and oilseeds produced throughout the year for consumption, and they prepare soft commodities for use lower down the value chain. Wheat, for example, must be milled into flour for bread or noodles, while soybeans must be crushed for oil or livestock feed.

Despite the fact that certain places produce greater output and traders decrease the supply-demand imbalance, doubling food production by 2050 will be a significant problem. To achieve a sustainable global food balance, businesses, and governments will need to collaborate to increase productivity, foster innovation, and improve supply chain integration. Farmers, riding firms, and other processing industries (especially Big Food) must first prepare for deforestation-free supply chains.

Deforestation not only causes rapid and irreversible biodiversity losses, but it is also the second largest source of CO_2 emissions after fossil fuels and has played a significant role in global warming—adding to the negative pressure on agriculture production that these forests were cleared for in the first place. "Sustainable intensification" refers to the requirement for farmers to grow on the land area where they are currently operating.

This suggests the use of precision instruments such as GPS fertilizer, an enhanced irrigation model, and a crop rotation schedule that is environmentally friendly.

The strategies listed above aid in the production of more crops, particularly in areas of Africa, Latin America, and Eastern Europe where yield gaps are substantial. These will help mitigate the negative environmental impact of over-stressing resources by reducing groundwater depletion and, as a result, the degradation of productive areas due to fertilizer overuse. A good long-term investment from commercial companies, as well as public spending, is required in the agriculture sector.

Because land investments have traditionally delivered strong returns, enhanced diversity, and surpassed inflation, top investors such as pension funds and sovereign wealth funds have already made large commitments to support global agricultural production and commerce in recent years. Nonetheless, compared to rich countries, developing countries agricultural investment plans have dropped over the previous 30 years, and significantly less money is spent on research and development. This leads to low productivity and sluggish production, and because financial sectors in developing nations provide fewer loans to farmers, both farmers and large enterprises are still constrained in their investment opportunities.

Governments must lower risks in order to attract greater funding and investment in agriculture. Regulators must modify laws that impede the financial inclusion of small, rural farmers. Quick loans and interest rate limitations, for example, are less common in bank lending. Furthermore, supportive policies, laws, and public infrastructure investments would aid in the development of a favorable investment climate for agriculture. The global food balance must be prioritized by policymakers, corporations, and consumers. International businesspeople involved in this supply chain must commit to and express the need for policy changes, as well as for developed countries to promote investment in regions with the greatest growth potential. It will determine our food security.

11.2 BACKGROUND

Precision agriculture is a term that is currently being used a lot in the business. It assists growers in better responding to variations within one field or series of fields in order to improve overall crop health and production. Digital agriculture tools are a collection of technology and software that give data to support precision farming decisions and, when utilized correctly, can help minimize waste, enhance revenues, and protect the environment. Retailers and growers are now using digital agricultural tools in their precision agriculture strategy, including field applications such as equipment and sensor platforms that regulate goods and purchases and may give real-time data. Platforms for software application development are likewise getting more popular, expanding the number of possibilities accessible. These digital agricultural solutions frequently do not collect and analyze data to aid retailers and growers in making a variety of crop production decisions. Some digital agriculture solutions can assist in increasing field output while also lowering environmental stress, which are discussed in the subsections.

11.2.1 MONITOR SOIL AND PLANT PARAMETERS

Growers can use digital agriculture tools to assess peak plant growth conditions and what nutrients their crops receive to assist them in meeting their yield targets on a field-by-field basis. Today's technology offers a wide range of alternatives. The placement of sensors around the fields to measure rainfall and soil conditions in these various places is one technique. This will aid in the reduction of VRT and the improvement of yields in major issue regions. These sensors work in conjunction with software tools that help producers take advantage of this technology. There are a variety of computer applications, software programs, and other digital agriculture technologies that can be used in conjunction with agricultural equipment or manually to record soil sample results, fertilizer inputs, rainfall, and other factors. Agronomists can use new technologies to better understand and improve their farming decisions for individual areas. Many electronic smart instruments are available that can be programmed to figure out how to operate a grower's equipment or can be manually programmed with field data.

11.2.2 AUTOMATE FIELD MANAGEMENT

If a grower uses sensor technology and systems, soil and plant species are frequently automatically optimized using these sensors and delivered from a choice network, which may help decide the most efficient watering and fertilization schedules for the actual crops. If a grower uses a digital tool like field view, they may track growth throughout the season using a range of resources available on the platform. Field health photography, for example, allows growers to track the status of the crop and support crop scouting and field management decisions as the season advances, allowing them to maintain the highest yield potential.

11.2.3 COLLECT REAL-TIME DATA

If a grower installs sensing devices across the field, it will provide continuous monitoring of the selected parameters as well as real-time data to help the producer make educated decisions during planting, the growing season, and harvest. Growers can use real-time data at important times to help them make decisions with digital agriculture solutions. If a grower has uploaded planting data or utilized the Field View Cab app during planting, for example, maps will be created in their account. These maps can subsequently be used during harvest, or they can be paired with harvest data to provide specific yield results by field, hybrid, or variety.

11.2.4 GET EXCELLENT RESULTS FROM LABOR AND RESOURCES

Growers and retailers are ready to employ technology to help maximize the benefits of crop nutrients, crop protection, and irrigation expenses by deploying autonomous sensors that notify growers when it's time to irrigate and fertilize their crops. Growers can develop a variety of seeding and fertility prescriptions (scripts) using digital agriculture tools like Climate Field View, incorporating nitrogen, phosphorus, potassium, and other critical nutrients. It is estimated that about half of all growers use agricultural technology, and this number is expected to continue to rise as agronomists and growers see the benefits of digital agriculture technologies and how they plan to enhance their results as part of their broader precision agriculture plans. There are also dealer choices for a few digital programs, allowing the agronomist to track results and analyze how crops performed across greater regions or under a variety of growing situations. Agriculture retailers have the opportunity to help their customers by providing local expertise on how the various digital agriculture tools and technologies available today can best be implemented within their geography and within their specific operations to help them improve their return on investment in the most effective ways.

Different technologies used in intelligent agriculture systems (IAS) are:

i. Sensing technology;
ii. Software applications;
iii. Communication systems;
iv. Positioning technologies;
v. Advanced information technology hardware;
vi. Data analytics.

11.3 PROPOSED MODEL

Smart farming is a new term for managing farms using contemporary information and communication technologies to increase the number and quality of crops while reducing the amount of human work required. The following are some of the technologies available to today's farmers:

- Soil, water, light, humidity, and temperature sensors;
- IoT platforms;
- Cellular, ZigBee communication technology;
- GPS, satellite services;

- Robotics, autonomous tractors, processing facilities;
- Data analytics solutions, etc.

11.3.1 SYSTEM DESIGN

To achieve the goals of the smart farming model, it will be necessary to develop an IoT-based smart farming system that will regulate high-voltage electrical devices such as water pumps, playhouse flaps, and other similar devices without requiring human intervention in environmental parameters such as soil moisture and temperature. For future data analysis, these parameters are saved in the internet cloud. For a more regulated environment, farming is done in playhouses. The suggested system will be made up of several levels, as shown in Figure 11.4. The sensor layer, middleware, communication layer, and cloud and application layer are the four modules:

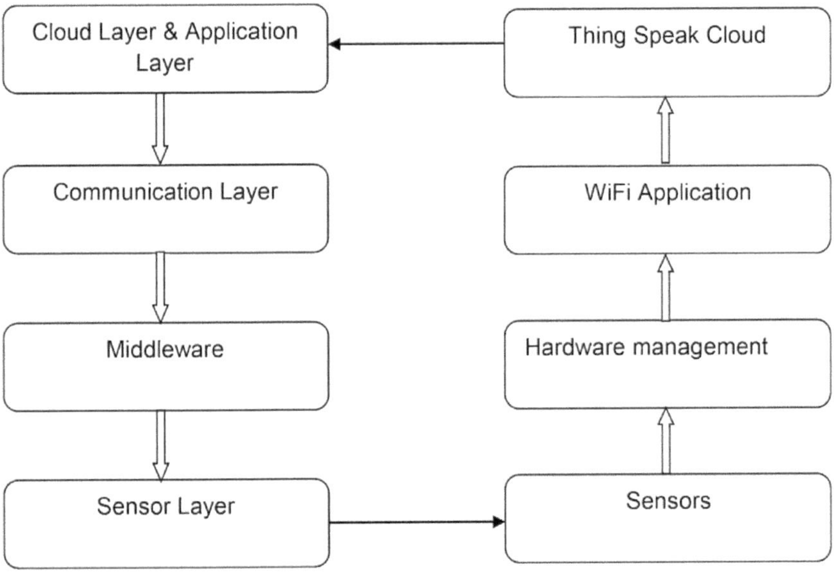

FIGURE 11.4 Smart farming technology stack.

1. **Sensor Layer:** This is the proposed system's initial layer. It is in charge of recording and monitoring various environmental

factors. This layer continuously senses or collects parameters from many types of sensors placed throughout the agricultural field. Two types of sensors were employed in this study: a soil moisture sensor to monitor soil humidity and a temperature sensor to measure temperature within poly homes. All of these sensors are connected to an Arduino-based microcontroller, and the microcontroller is connected to sensors to construct basic IoT objects that are used in agriculture.

2. **Middleware:** The suggested system's second layer is needed to automate the farming operation and control the actuators. It will be made for a microcontroller. The microcontroller receives the sensed values as input and acts accordingly based on the measured values of several monitoring field parameters. This layer keeps a close eye on the soil temperature and moisture level, as these two factors have a direct impact on crop output and subsequent actions.

    - If the moisture level in the soil does not reach the saturation level, the microcontroller will engage the pump machine to water the sector, as the insufficient moisture content in the soil reduces crop output. The soil moisture content threshold varies depending on the kind of soil [12]. According to Ref. [12], the appropriate moisture content threshold values for different types of soil where irrigation takes place are listed in Table 11.1. A threshold of 15% soil moisture content is used in the proposed method. When the moisture level hits the limit, the pump will shut down automatically, saving you money on electricity.
    - If the temperature level exceeds the edge value, the microcontroller will open the poly house flap. The proposed system uses a temperature threshold of 40°C. Temperature rises shorten crop life and disrupt the delicate balance between crops and pests. It also boosts crop respiration rates while decreasing fertilizer efficiency. Apart from operating the actuators, the microcontroller communicates the detected data from the sector to the Thing Speak cloud via a gateway.

3. **Communication Layer:** Because Wi-Fi has an advantage over Bluetooth, the microcontroller connects with the gateway wirelessly in this layer. Bluetooth offers a shorter range of connections

than Wi-Fi, and the gateway may be located outside of the monitoring area. Because of the extensive wiring, Ethernet-based communication is avoided. Here, a microcontroller is equipped with sensors that are placed across the monitoring field, delivering the measured moisture in the soil and temperature to the cloud via a gateway. On the gateway, an IP-based protocol is in use. The microcontroller sends an HTTP request to the Thing Speak cloud, requesting that the sensed value be sent to the appropriate channel.

4. **Cloud Computing:** It is a new technology that has the potential to be useful in smart farming. This approach proposes using a cloud computing platform to record several aspects of agricultural field data. Different channels are formed throughout this layer, each of which corresponds to a specific parameter field within the Thing Speak cloud for storing field data (temperature, soil moisture). Through a communication protocol, the microcontroller sends the sensed real-time data to the appropriate channel on a regular basis. These data (soil moisture value, temperature value) are graphed over time and can be used for further investigation. Agricultural field conditions (temperature, soil moisture) are frequently checked remotely using the Thing Speak online service's graph.

Farming-related applications are frequently developed and deployed on the cloud, where they can be used by farmers (Figure 11.5).

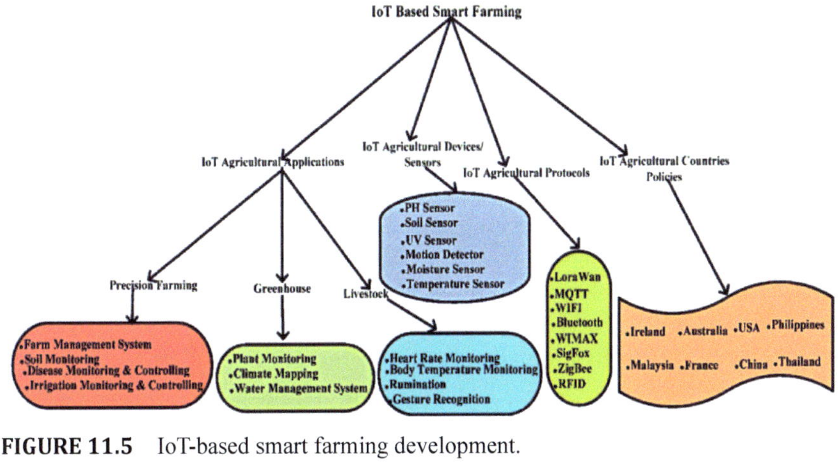

FIGURE 11.5 IoT-based smart farming development.

Source: Adapted from [11]. https://creativecommons.org/licenses/by/4.0/

TABLE 11.1 Soil Type and Moisture Content

Type of Soil	Texture and Moisture Content in Soil (%)
Simple	7.0
Loamy	12.0
Loam (sandy)	15.0
Silt	20.0
Loam	23.0
Silty (clay)	28.0
Clay	27.0
Sandy clay loam	24.0
Simple sandy clay	22.0
Silty clay	30.0
Clay	31.0

Various equipment is being employed to realize the envisioned system. All sensors are connected to the Arduino UNO board, which is often used as a microcontroller. The LM35 is frequently used as a temperature sensor, whereas the VL95 is frequently used as a soil moisture sensor. To regulate high-voltage devices, stepper motors and fans are frequently linked to Arduino UNO boards using 6-pin relays. The detected data is supplied into the Arduino board's middleware, which supports these values and controls the actuator (Controlling Pump, Fan). Apart from operating actuators automatically, the Arduino IDE is used to create middleware. The Arduino board also sends the measured parameters to the Cloud platform. For storing environmental parameters, Arduino boards communicate wirelessly with the Thing Speak cloud through a router. During this model, Wi-Fi-based communication is used. As a Wi-Fi module, the ESP8266 module is used. To communicate with the Cloud, the ESP8266 module connects to a specialized gateway device with internet access. The proposed system has collected the temperature and moisture of the soil from the monitoring field to support the experimental setup. These observed data are then plotted in the cloud-based Thing Speak web service every 15 seconds, as Thing Speak requires a 15-second delay between periodic updates. Different levels of soil moisture and temperature were measured and supported predetermined threshold values of soil moisture and temperature, and the Arduino board manages the high-voltage farming equipment without human interaction, based on the above-mentioned system design. This technology provides continuous field monitoring and triggers the appropriate occurrences corresponding with the need in the absence of a human in the agriculture field. It reduces the amount of human work and the cost of farming to a certain level. In order to test the proposed technology in the various soil texture environments listed in Table 11.1, the edge value of soil moisture and temperature must fluctuate, and this can be incorporated into the middleware by manually changing it.

The proposed model may need above all sensors and devices (Figure 11.6) that are to be used in a real-time environment for framework design. The sensors and devices can be attached and integrated with other modules also, like data analyses, processes, visualization, and prediction (Figure 11.7). The analysis and prediction module of the framework will use some cloud applications to manage the data generated by sensors as well as it can help to predict the weather and other parameters so that resources can be saved by the farmer as well as government authorities involved like the Electricity Distribution Company.

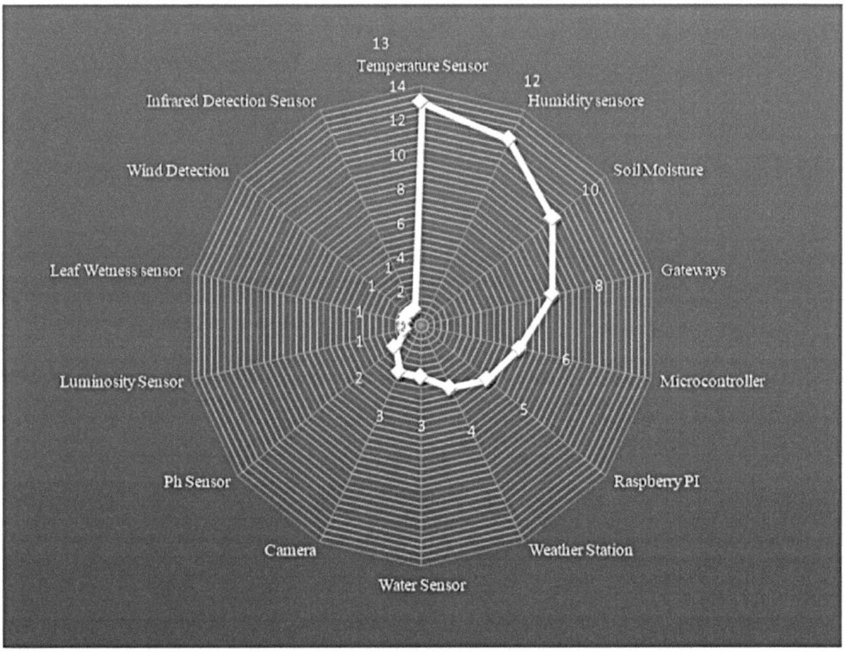

FIGURE 11.6 Sensors and device distribution.

Source: Reprinted from [11]. https://creativecommons.org/licenses/by/4.0/

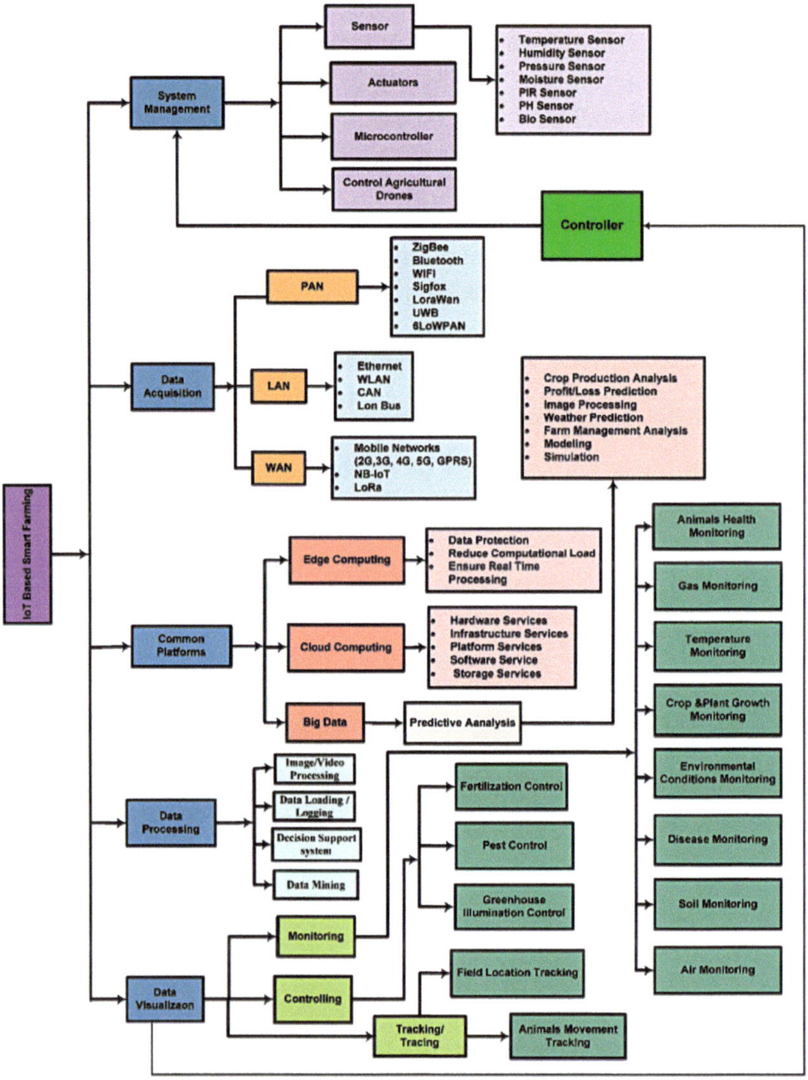

FIGURE 11.7 Smart farming modern design and application.

Source: Reprinted from [11]. https://creativecommons.org/licenses/by/4.0/

11.4 RESULTS

The proposed IoT framework for agriculture is sort of interesting and a recently developing area, and therefore the results of the proposed system are often compared with an existing system with previous IoT automated agriculture works. This proposed model has worked with issues and challenges related to the implementation of IoT applications of previous works like security, cost, reliability, scalability, localization, and interoperability.

Users encounter numerous issues as a result of insufficient security, including knowledge loss and other on-field parameters. Physical interference, such as attacks by animals and predators or changes in physical address, puts IoT devices in the agriculture field in jeopardy. Furthermore, sophisticated and complicated algorithms are difficult to implement due to low energy consumption and limited memory. Hijacking assaults, session hijacking, database difficulties, and denial of service attacks are all major security threats to cloud infrastructure.

Several cost-related challenges exist when using IoT in agriculture, such as setup and running costs. Hardware costs like IoT devices/sensors, a base station (BS) infrastructure, and gateways are included in the setup costs. Furthermore, ongoing costs include a continuous subscription for IoT device administration, data sharing, and other services, as well as centralized services that provide information/data collecting. The major challenge for farmers living in rural areas is a lack of understanding of technology. This is a prevalent problem in underdeveloped countries since the majority of farmers are illiterate.

The adoption of IoT in agriculture could be difficult due to the significant expenditure necessary in farmer training prior to the deployment of IoT infrastructure. In the realm of agriculture, IoT devices are installed in an open environment, where harsh external conditions might cause communication failure and, as a result, sensor humiliation. As a result, it's critical to ensure the physical safety of deployed IoT devices/sensors in order to protect them from harsh weather conditions. In the agricultural sector, a high number of IoT devices and sensors are deployed, necessitating the use of an intelligent IoT management system for the identification and control of every node.

When it comes to deploying devices/sensors, there are numerous variables to consider. Without deploying new devices with extra configuration, such devices should be able to provide functionality and support to the rest of the planet. Furthermore, the simplest deployment position should be chosen so that devices can connect and exchange information without hindrance. To interoperate, billions of IoT devices, standards, and protocols are required. Semantic, syntactic, technical, and organizational policies all play a role in interoperability.

Semantic interoperability refers to the ability to alter the understanding of content communicated among humans. Data formats such as JavaScript object notation (JSON), data interchanged electronically, extensible terminology (XML), and variables separated by a comma are considered to have syntactical interoperability. The occurrence of infrastructure, protocols, and hardware/software components that enable the connection of IoT devices is referred to as technical interoperability. Interoperability policies are defined to be policies for properly communicating and transferring data across multiple geographic regions and infrastructures. In contrast, three strategies are offered for achieving interoperability: (i) two open and closed standards; (ii) partnership among services and merchandise providers; and (iii) mediator and adaptor services. More research is expected to be done in order to achieve high interoperability among multiple IoT devices.

11.5 CONCLUSION

The chapter recommends combining the most recent innovation in the field of autonomous agriculture with current tactics to indicate on/off for water systems, resulting in simple profitable, and temperate trimming. Some interesting technology is displayed, transforming the idea of an agricultural field into a product that manages sensor data via cloud administrations. Many points of interest have been started to use sensors that employment consequently improving. This concept of modernization of farming is cheap and operable for farmers and a requirement of today's life. The thought of modernization is even developing and continuously growing as a change in digital circuit improvements also as technology to handle the system cannot stop this research. The predictive model of the framework, which involves data analysis, can improve the functionality of the model, and with the assistance of this, we will save a lot of resources like electricity and manual human resources (HRs) and can make strategies for disaster and may be functional for farmers. If all the parameters for irrigation with smart IoT devices are set for a machine learning model, this system can improve the agriculture system of a rustic, which is productive and effective.

KEYWORDS

- **extensible terminology**
- **intelligent agriculture system**
- **Internet of things**
- **JavaScript object notation**
- **mediator and adaptor services**
- **monitor soil**

REFERENCES

1. Abhijit, P., Mohammad, A., Md. Jainal, A., Karl, A., Rashed, M., & Mohammad, S. H., (2019). IoT-based smart system to support agricultural parameters: A case study. *The 6 International Symposium on Emerging Inter-Networks, Communication and Mobility (EICM) 2019* (648–653). Halifax, Canada.
2. Vinayak, N. M., & Pooja, K. A. (2017). Role of IoT in agriculture. *IOSR Journal of Computer Engineering (IOSR-JCE).* e-ISSN: 2278-0661, p-ISSN: 2278-8727 PP 56-5.
3. Muhammad, A., Ammad-Uddin, M., Zubair, S., Ali, M., & El-Hadi, M. A., (2019). *Internet-of-Things (IoT)-Based Smart Agriculture: Toward Making the Fields Talk* (Vol. 7). Special section on new technologies for smart farming 4.0: Research challenges and opportunities, IEEE Access.
4. Umair, K., Muhammad, M. I., Meshrif, A., Mohammed A. Al. G., Muhammad, R., & Sultan, H. A., (2020). Text data security and privacy in the internet of things: Threats, challenges, and future directions. *Wireless Communications and Mobile Computing, 2020.*
5. Muthunoori, N., & Munaswamy, P., (2019). Smart agriculture system using IoT technology. *International Journal of Recent Technology and Engineering (IJRTE), 7*(5). ISSN: 2277-3878.
6. Ritika, S., Vandana, S., Vishal, J., & Sumit, R., (2020). A research paper on smart agriculture using IoT. *International Research Journal of Engineering and Technology (IRJET), 7*(07).
7. Partha, P. R., (2017). Internet of things for smart agriculture: Technologies, practices and future direction. *Journal of Ambient Intelligence and Smart Environments, 9*, 395–420.
8. Md Ashifuddin, M., & Zeenat, R., (2019). IoT-based intelligent agriculture field monitoring System, *Second International Conference on Advanced Computational and Communication Paradigms (ICACCP-2019).*

9. https://voxeu.org/article/demand-agricultural-commodities-grow-more-slowly-next-decade (accessed on 21 December 2022).
10. Saiz-Rubio, V., & Rovira-Más, F., (2020). From smart farming towards agriculture 5.0: A review on crop data management. *Agronomy, 10*, 207.
11. Muhammad, S. F., Shamyla, R., Adnan, A., Tariq, U., & Yousaf, B. Z., (2020). Role of IoT technology in agriculture: A systematic literature review. *Electronics, 9*, 319.
12. Zhang, Y., (2019). The role of precision agriculture. *Resource, 19*, 9.
13. Himesh, S., (2018). Digital revolution and big data: A new revolution in agriculture. *CAB Rev., 13*, 1–7.
14. Agrawal, S., & Das, M. L., (2011). Internet of things—A paradigm shift of future internet applications. In: *International Conference on Current Trends in Technology* (pp. 1–7). IEEE.
15. Abdurrahman, M. A., Gebru, G. M., & Bezabih, T. T., (2015). Sensor-based automatic irrigation management system. In: *International Journal of Computer and Information Technology (ISSN: 2279 – 0764)* (Vol. 04, No. 03).
16. Schimmelpfennig, D., (2016). *Farm Profits and Adoption of Precision Agriculture, 217*, 1–46. USDA.
17. Grand View Research, (2019). *Precision Farming Market Analysis* (pp. 1–58). Estimates and trend analysis; Grand View Research Inc.: San Francisco, CA, USA.
18. Díez, C., (2017). Hacia una agricultura inteligente (towards and intelligent agriculture). *Cuaderno de Campo (Field Notebook), 60*, 4–11.
19. Accenture Digital. *Digital Agriculture: Improving Profitability.* Available online: https://www.accenture.com/_acnmedia/accenture/conversion-assets/dotcom/documents/global/pdf/digital_3/accenture-digital-agriculture-point-of-view.pdf (accessed on 21 December 2022).
20. CEMA. *Digital Farming: What Does It Really Mean?* Available online: http://www.cema-agri.org/publication/digital-farming-what-does-it-really-mean (accessed on 21 December 2022).
21. Nierenberg, D. *Agriculture Needs to Attract More Young People*. Available online: http://www.gainhealth.org/knowledge-centre/worlds-farmers-age-new-blood-needed (accessed on 21 December 2022).
22. European Commission, (2012). *Generational Renewal in EU Agriculture: Statistical Background* (pp. 1–10). DG Agriculture & Rural Development: Economic analysis of EU agriculture unit: Brussels, Belgium.
23. Paneva, V. *Generational Renewal*. Available online: https://enrd.ec.europa.eu/enrd-thematic-work/generational-renewal_en (accessed on 21 December 2022).
24. Alpha Brown. *What is IoT in Agriculture? Farmers Aren't Quite Sure Despite $4bn US Opportunity—Report*. Available online: https://agfundernews.com/iot-agriculture-farmers-arent-quite-sure-despite-4bn-usopportunity.html (accessed on 21 December 2022).
25. https://inthefurrow.com/ (accessed on 21 December 2022).

CHAPTER 12

Synthesis and Fabrication of a Nanosensor Device for Monitoring Nutrient Levels in Aeroponic Agricultural Farming

D. GAJALAKSHMI

Assistant Professor (SG), Department of Chemistry, University College of Engineering (A Constituent College of Anna University, Chennai), Villupuram, Tamil Nadu, India

ABSTRACT

Nanotechnology had substantial applications in recent years. Nanotechnology will involve a wide range of academic disciplines, demanding inter- and transdisciplinary partnerships. Nanotechnology has the ability to produce novel sensors, materials, and equipment on the nanoscale level. Nanoparticles have a wide range of applications due to their non-toxicity, non-immunogenicity, and ability to be customized for specific objectives. By adapting to any environment, nanosensors can be used to tackle a range of problems and treat numerous diseases. Nanosensors can be created utilizing a variety of methods, including the creation and manufacturing of nanoparticles using environmental conditions. This paper examines breakthroughs in nanosensor development for nutrient-level detection to support aeroponic agriculture.

Intelligent Sensor Node-Based Systems: Applications in Engineering and Science, Anamika Ahirwar, Piyush Kumar Shukla, Prashant Kumar Shukla, and Ruby Bhatt (Eds.)

12.1 INTRODUCTION

Daily, agricultural land is being depleted to fulfill the demands of India's growing population. Farmers face enormous difficulties in providing food for such a large population. Due to the natural growing process, cultivated crops are insufficient. Farmers use fertilizers to boost crop yields. Without a doubt, chemicals increase crop productivity.

Alternatively, using fertilizers degrades the nutritional content of food. In addition, these compounds can cause an excessive number of skin illnesses and disturb the entire food chain. The aeroponic approach can be utilized to reduce ecological disruption and promote a healthy lifestyle to address this issue. Aeroponics can be defined as growing plants without requiring soil or water as a medium, employing constant conditions such as temperature, humidity, pH, and the electrical conductivity of the nutrient solution essential for plant growth.

The advantage of aeroponics is that it promotes pink plant growth and the production of nutritious fruits by utilizing fewer nutrients and water from the environment/soil. This technology can produce fresh and healthy fruits on a perpetual basis. In recent decades, aeroponics has been utilized to cultivate and produce potatoes and potato seeds. Aeroponics has been actively employed in recent decades to grow disease-free potato seeds, and it is envisioned to create a pest-free growth environment eventually. Aeroponic farming is a way of growing plants in an air or water environment without soil or aggregate media. Aeroponics creates a closed atmosphere and water/nutrient environment that promotes rapid plant development without soil or medium. Aeroponic farming utilizes nitrogen, phosphorus, potassium, and secondary nutrients such as calcium, magnesium, and sulfur. In addition, aeroponic farming requires the following micronutrients: "iron, zinc, molybdenum, manganese, boron, copper, cobalt, and chlorine." The primary advantages of aeroponic farming are as follows:

- Energy consumption is lower than in traditional agriculture;
- Water use is reduced;
- Less upkeep is required because air functions as a medium for plant growth;
- Disease free cultivation;
- Increased exposure to the air;
- Eco-friendly, as no runoff from fertilized soil occurs;
- A negligible percentage of water is lost via evaporation;
- Beneficial for places afflicted by drought.

The nutrient solution, which contained all of the needed macro- and micro-nutrients, was periodically sprayed into the root zone as a fine mist to keep the plant nutrient saturated. The pH of the nutritional solution was maintained between 5.5 and 6.5, and the electrical conductance was observed between 2 and 2.5 mS/cm. The nutrient solution was replenished every two or three weeks to ensure enough nutrient levels and to maintain the proper pH level. This necessitates using an appropriate substance to monitor the medium's nutrition content. To ensure the most efficient use of aeroponic growing methods for plant production, it is critical to have an intelligent screening and control system for water and fertilizer delivery. Unfortunately, there has been no provision in the aeroponics equipment for a sensor to check the nutrient level. Thus, the primary purpose is to produce a material capable of sensing the level of nutrients in a medium. The purpose of this chapter is to demonstrate the appropriateness of aeroponics with a sensor in drought-stricken areas to tackle water scarcity.

12.1.1 DESCRIPTION OF THE AEROPONIC SYSTEM

Aeroponic growth systems are optimal for growing food in a range of unusual locations – both indoors and outdoors. Aeroponic farming requires a minimum of 200 square feet, a freshwater source, an electrical vent, and sunlight.

A proficient aeroponics system depends on a hassled water tank that can hold 60 to 90 psi and a high-quality mister competent in producing the tiniest mist of moisture conceivable.

12.1.2 WHAT TYPES OF PLANTS CAN AEROPONICS GROW?

Aeroponics is a more effective method of growing tuber crops than hydroponics, as the tubers have sufficient area to grow and are easily harvestable. Vegetable crops can be grown, but they have more complex nutrient requirements. In addition, fruit-yielding shrubs and trees are not viable in aeroponics systems due to their size.

12.1.3 AEROPONICS' ADVANTAGES

1. **Crop Yields can be Increased Significantly in a Limited Space:** A single Tower Farms unit is less than 6 square feet in size and

comprises stackable parts capable of housing four plants each (for example, a 10-section-tall team may hold 40 plants). As a result, Aeroponics farms can generate up to 10 times the amount of food grown on the same land area as a regular farm of comparable size without incurring additional land costs.

2. **Contributes to up to 98% of Water Conservation:** Tower Farms systems employ closed-loop water recycling, consuming up to 98% less water than conventional farms, which is critical for farmers in drought-prone areas.
3. **Farming-Related Problems can be Avoided:** By growing crops away from the soil, the risk of soil-borne pests and plant diseases can be avoided. In addition, because each tower was developed independently, any issues that arise may be isolated and resolved without jeopardizing the entire farm.
4. **Reduces the Number of Agricultural Steps:** Farming is a physically demanding occupation. Tower Farms simplifies operations by automating feed and water cycles, eradicating weeds, and lowering insect threats.
5. **Spending More Time in the Air than Water:** Aeroponic plants spend 99.98% of their time in the air, and only 0.020% are directly touched with a hydro-atomized nutrient solution. The time spent without water enables the roots to get oxygen more efficiently. Additionally, the hydro-atomized mist significantly adds to the roots' adequate oxygenation. Because fewer nutrients are required in volume, the amount of nutrients necessary for plant development is likewise lowered.

   The reduced nutrient intake considerably reduces the water consumed, requiring less space. As volume and space consumption is reduced, this paves the way for the development of lighter aeroponic systems.
6. **Low Water Consumption:** Because low water consumption reduces effluent produced by plants, water treatment for reusing the water significantly decreases. Reduced root-to-root contact between plants also helps to minimize disease transmission between plants.

7. **Increased Control Over the Environment in which the Plant Grows:** Aeroponics allows for the administration of various nutrient solutions directly to the roots without the need to rinse out any previous solution or matrix in which the roots were immersed. This enables the investigation of the effects of various fertilizer treatments on the roots of the plant species being studied. In addition, aeroponics, like other nutrient delivery systems, is better suited for a broader range of growing conditions. For example, the nutrient spray interval and duration can be precisely tuned to meet the needs of a particular plant species.
8. **Convenient to Use:** Aeroponic systems are designed so that they are simple to work with plants. The plants are isolated due to their suspension in the air and their roots being free of matrix. As a result, individual harvesting plants are relatively simple and rapid. Similarly, removing any plant that appears to be infected with a virus is straightforward and poses little risk of uprooting or infecting neighboring plants.
9. **Reasonably Priced:** Aeroponic systems are more affordable than other systems. In addition, due to the reduced volume of solution throughput, less water, and fewer nutrients are required than in conventional nutrient delivery systems.
10. **Seed Stocks Free of Pathogens were Created:** Aeroponics can help minimize disease spread by pathogen-infected plants. As previously noted, this is owing to the plants' separation and the absence of a shared growth substrate. Additionally, aeroponics may be an ideal technology for developing pathogen-free seed stocks due to the enclosed, regulated environment. Along with the isolation mentioned above of plants, the growing chamber's enclosure assists in preventing initial contamination by pathogens introduced from the outside environment and minimizes the transfer of any diseases that may exist between plants.

12.1.4 NEGATIVE ASPECTS

1. Aeroponics systems require a certain level of skill to operate efficiently. The vital metals absorption in the water must be strictly controlled, and even a minor malfunction in the operation of the

equipment might result in crop loss. For example, if the misters do not spray every few minutes, those dangling roots will quickly desiccate. Additionally, the misters must be cleaned regularly to prevent clogging from mineral deposits in the water.

2. Aeroponic systems require electrical power to pump water through tiny misting devices, a significant environmental drawback. However, they are suitable for use in a greenhouse that receives natural light. Renewable energy or other forms of alternative energy can be employed to evade this drawback.
3. The primary disadvantage of aeroponic farming is maintaining a predetermined nutrition level. This rate of nutrient absorption varies amongst plants and is also size dependent. If the nutrient level is not kept within the required limits, it clogs the mister, interfering with the continuous mist spray. This causes the plants to wither and decreases crop output. Another significant effect is that if the nutrient level is left unregulated following plant absorption in a large-scale aeroponic farm, it will result in soil pollution when the nutrients are drained into the soil. Drained water containing nutrients from the soil eventually pollutes the soil. Additionally, it will contribute to Eutrophication via surface runoff water.

The primary purpose of this research is to construct a device for monitoring nutrient levels in the medium used in aeroponic farming. This would ensure maximum nutrient absorption from the medium, leaving the leftover water free of fertilizer that may be released into any water bodies, effectively assisting in the prevention of:

- Eutrophication;
- Pollution of the soil;
- Pollution of the water;
- Especially beneficial in places prone to drought.

12.2 REVIEW OF LITERATURE

- Martha M. Vaughan et al. [1] discovered that the aeroponic culture of *Arabidopsis* and the generalist herbivore *Bradysia* (*B. coprophilia* and *B. impatiens*) could be utilized to explore subsurface plant defense responses to herbivorous attacks such as secondary metabolite production.

- Frederic F. Souret & Pamela J. Weathers [2] investigated saffron production in aeroponic, hydroponic, and soil mediums. The development of stigmas and the concentration of important saffron elements in the stigmas were identical in all three growing techniques, indicating that saffron bulbs might be exploited as a sustainable source of saffron-derived medicinal chemicals.
- Margaret Chiipanthenga et al. [3] found that traditional potato seed manufacturing is ineffective at preventing or lowering pathogen structure, resulting in lower-quality potato seeds and crop yields. It is observed that aeroponics' capacity to appreciate plant tissue culture enabled it to pioneer the production of potato seeds in underdeveloped countries' agricultural systems.

The preceding paragraphs detail the experimental inquiry into the growth of several plants using the aeroponics method and a little investigation into the plants' nutrient consumption from the medium. This chapter aims to synthesize, analyze, and print a metal nanoparticle on conducting media. An amplifier circuit is used to verify the printed electrodes' suitability for sensor use. The emphasis here would be on nutrient uptake monitoring in the aeroponics farming system. This would indicate the maximum and minimum nutritional levels via the sensor display.

Aeroponic farming research is being conducted in only a few laboratories in India. Lakkireddy et al. [5] investigated the role of hydroponics and aeroponics in commercial food production using soilless culture.

Tk Bag et al. [6] employed aeroponics to grow high-quality potatoes in the Northeastern Himalayas. They demonstrated that an aeroponic system could be used successfully to improve the yield of disease-free, high-quality potato seeds.

12.3 OBJECTIVES

Food contamination can occur in a variety of ways. Food products can get contaminated at any point along the supply chain, including during manufacture, processing, shipping, and distribution. Food contamination can be classed generically as biological, physical, or chemical. When food is contaminated with pathogenic bacteria such as *Salmonella* spp. or poisonous organisms such as *Clostridium botulinum*, viruses such as norovirus cause food poisoning and deterioration. This is referred to as biological food poisoning. When food is contaminated with heavy metals, dioxins, pesticides, or radionuclides, chemical contamination occurs. Chemical food poisoning is the term used to describe food poisoning caused by chemicals. People are seeking alternate food production methods as they become more conscious of the chemicals and pollutants included in food goods. Aeroponic farming is the best non-chemical method for food cultivation. Aeroponic farming is ideal for an area that is unsuitable for plant development. This method of farming consumes very little water. The growth media for plants is reusable and recyclable. Growing a tiny aeroponics garden is becoming popular for personal delight and domestic veggie consumption.

With all its advantages, the chapter's primary objective is that blossoming aeroponic farming requires a suitable monitoring mechanism to maintain the medium's nutrient level. Therefore, designing an inventive way to monitor the aeroponic system's nutrition content. Unfortunately, there is no effective mechanism for monitoring the nutrient limits in an aeroponic medium. This chapter aims to discuss the synthesis of anode and cathode materials. Validation of the synthesized substance is accomplished by using appropriate characterization techniques. Following characterization, the materials are printed on substrates such as glass, semiconductor materials, and even paper. Additionally, this chapter discusses the manufacture of sensors employing electronic components such as the AD620, ARM Processor, and others to display the plant's nutrient level. The chapter's primary aims are as follows:

- To synthesize the components;
- To characterize materials using a variety of approaches;
- To print on a suitable substrate, the synthesized substance;
- To create the printed material by the use of electronic components.

The chapter is remarkable in that it synthesizes the material and fabricates it as a device. As it is known in India, Aeroponic farming uses a limited number of sensors or none at all to monitor nutrient levels. While aeroponic farming is a well-known method of agriculture in India, it lacks a sensor that measures nutrient levels. Thus, this chapter will address this void by synthesizing, characterizing, and fabricating a Nano-Sensor for use as a nutrient level monitor. The chapter's overall structure is depicted in Figure 12.1.

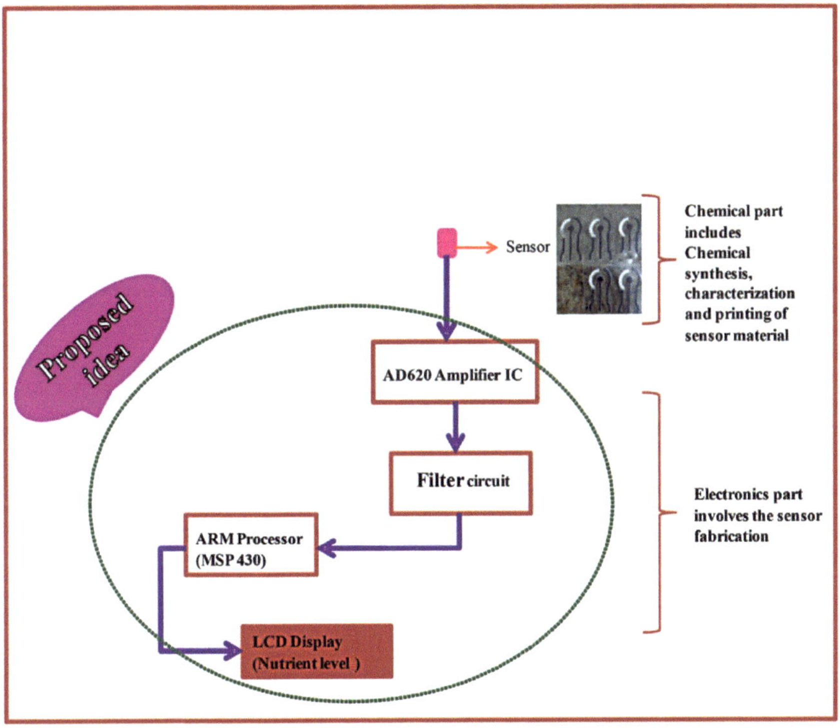

FIGURE 12.1 Scheme of the work.

12.4 TECHNIQUES

This chapter's research approach is divided into three sections:

i. Synthesis;
ii. Characterization; and
iii. Fabrication.

12.4.1 SYNTHESIS AND ANALYSIS

To obtain a flocculate, metal acetate (0.1 mol) and PVP (0.02 mol) were dissolved in ethylene glycol and refluxed for two hours at 195°C. Centrifugation is used to collect the flocculate, then washed with deionized water and ethanol. The precipitate is dried in a vacuum oven at 80°C for 2 hours and then analyzed with TG/DTA to determine the crystallization temperature in its entirety. After precipitation, the precipitate is calcined in the air at 500°C for two hours to obtain the required metal nanoparticle. Finally, the residue is characterized using techniques such as SEM, TEM, FESEM, CV analysis, FTIR, FT-Raman, and UV, among others.

12.4.2 TECHNIQUES FOR FABRICATING ORGANIC DEVICES

One critical way to fabricate organic electrical devices is by using classic vacuum and lithography techniques. On the other hand, traditional methods require more expensive manufacturing processes, such as chemical or physical vapor deposition and plasma etching, which demand higher temperatures and vacuum. However, solution-based techniques are both affordable and environmentally friendly. Therefore, a new fabrication technique based on solution processability has gained prominence in recent years. The inkjet printing technique is one of the essential methods.

Inkjet printing is a fast-increasing technique that allows for the delivery of a small amount of material in the form of ink to the desired spot on a substrate.

This printing technique may be used to make exact graphical patterns and printed electrical devices such as circuit boards, solar cells, PLEDs, and RFIDs.

It is also used in the pharmaceutical business to administer medications. Additionally, it is utilized in the biomedical area to print DNA microarrays and stem cells. Figure 12.2 is a flow chart illustrating several inkjet printing procedures. While there are other printing technologies available, we chose continuous inkjet printing. This process is used to fabricate a nano-sensor, as illustrated in Figure 12.3. A typical inkjet-printed nanosensor is depicted in Figure 12.4.

12.5 IMPACT ANALYSIS AND RISK ANALYSIS ON THE ENVIRONMENT

The device would not be environmentally dangerous. On the contrary, it contributes to maintaining water and soil quality standards. For example, in aeroponic farming, the nutrient medium is supplemented with nutrients and a buffer solution. However, after two or three weeks of plant absorption, this solution will pollute the soil and even the water. The use of a

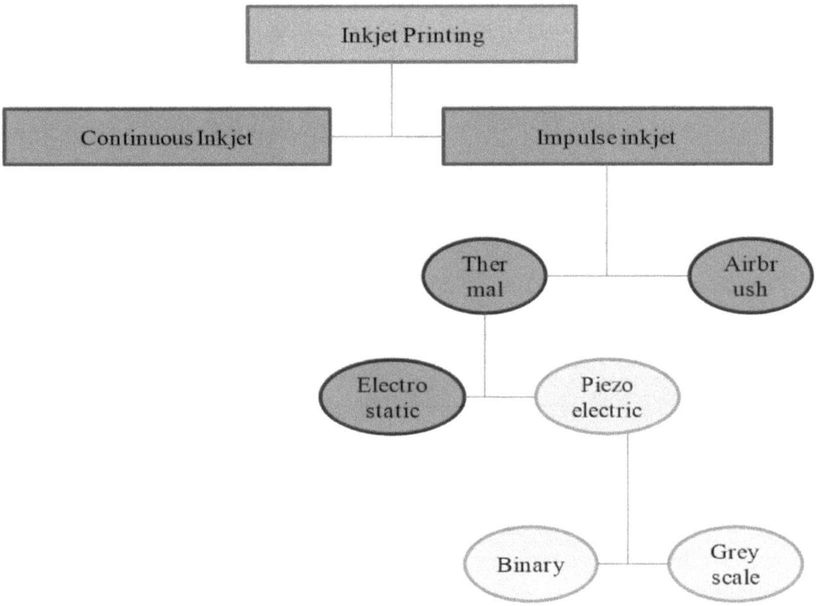

FIGURE 12.2 Types of inkjet printing.

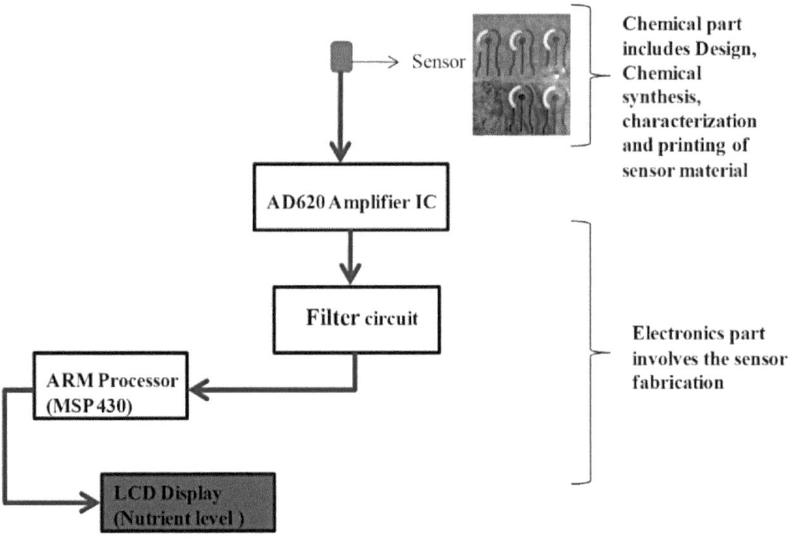

FIGURE 12.3 Sensor fabrication.

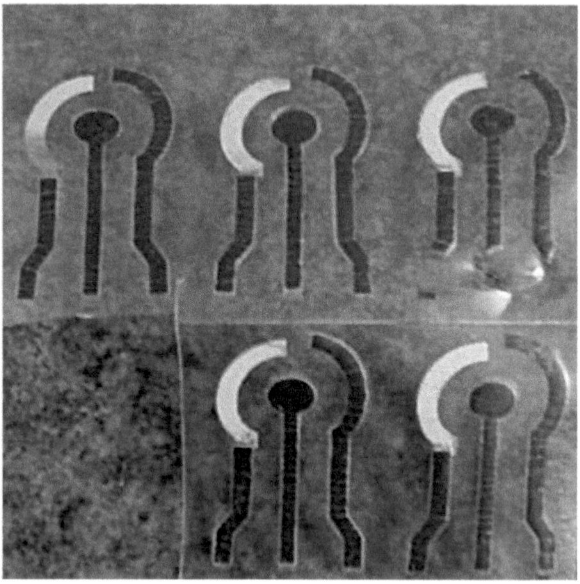

FIGURE 12.4 Inkjet printed nano-sensor.

sensor gadget eliminates this act of soil and water pollution. As a result, there is no risk associated with this work [7–16].

12.6 CONCLUSIONS

This research aims to improve aeroponic crop growth without damaging the land or water through the use of a nano-sensor device. The nano-sensor gadget is manufactured using an inkjet printing technique. This inkjet-printed nano-sensor gadget can effectively monitor the nutrient solution to avoid Eutrophication. This research will assist people in obtaining nutritious food and living a disease-free, happy life by eating pest-free crops and vegetables cultivated nearby, which is a basic need of all human beings.

KEYWORDS

- **aeroponic systems**
- **Arabidopsis**
- ***Clostridium botulinum***
- **fertilizers**
- **nano-sensor**
- **plant growth**

REFERENCES

1. Martha, M. V., Dorothea, T., & James, G. T., (2011). An aeroponic culture system for the study of root herbivory on *Arabidopsis thaliana*. *Plant Methods, 7*(5), 1–10.
2. Frederic, F. S., & Pamela, J. W., (2000). The growth of saffron in aeroponics and hydroponics. *Journal of Herbs, Spices and Medicinal Plants, 3*, 25–35.
3. Margaret, C., Moses, M., Paul, D., & Joyce, N., (2012). Potential of aeroponics system in the production of quality potato (*Solanum tuberosum* L.) seed in developing countries. *African Journal of Biotechnology, 11*(17), 3993–3999.
4. Ali, A., (2017). Hydroponics, aeroponics, and aquaponics as compared with conventional farming. *American Scientific Research Journal for Engineering, Technology, and Sciences (ASRJETS), 27*(1), 247–255.
5. Lakkireddy, K. K. R., Kasturi, K., & Sambasivarao, K. R. S., (2012). Role of hydroponics and aeroponics in soilless culture in commercial food production. *Journal of Agricultural Science and Technology, 1*(1), 26–35.
6. Bag, T. K., Srivastava, A. K., Yadav, S. K., Gurjar, M. S., Diendgoh, L. C., Rai, R., & Sukhwinder Singh, (2015). Potato (solanum Tuberosum) aeroponics for quality seed production in north Eastern Himalayan region of India. *Indian journal of Agricultural sciences, 85*(10), 1360–1364.
7. Chiipanthenga, M., (2012). Potential of aeroponics system in the production of quality potato (*Solanum tuberosum* L.) seed in developing countries. *African Journal of Biotechnology*, *11*(17), 3993–3999.
8. FAO Statistical Yearbook, (2013). *World Food and Agriculture*. http://www.fao.org/docrep/018/i3107e/i3107e.PDF (accessed on 21 December 2022).
9. Howard, M. R., (2012). Plant culture. *Hydroponic Food Production* (7 revised edn.).
10. Farran, I., & Mingo-Castel, A. M., (2006). Potato mini tuber production using aeroponics: Effect of plant density and harvesting intervals. *American Journal of Potato Research, 83*(1), 47–53.

11. Idris, I., & Muhammad, I. S., (2012). Monitoring and control of the aeroponic growing system for potato production. *Control, Systems & Industrial Informatics (ICCSII), 2012 IEEE Conference. IEEE.*
12. Jayapalan, M., & Kumari, S. N. P., (2001). Constraints in the cultivation of bitter gourd (*Momordica charantia* L.). *Journal of Tropical Agriculture, 39*(1).
13. King, M., & Zhu, B., (1998). In: Tang, S., & King, M., (eds.), *Gaming Strategies in Path Planning to the West* (Vol. II, pp. 158–176). Xian: Jiaoda Press.
14. Nugaliyadde, M. M., (2005). An aeroponic system for the production of pre-basic seed potato. *Annals of the Sri Lanka Department of Agriculture, 7*, 199–288.
15. Ritter, E., (2001). Comparison of hydroponic and aeroponic cultivation systems for the production of potato mini tubers. *Potato Research, 44*(2), 127–135.
16. Rolot, J. L., & Seutin, H., (1999). Soilless production of potato minitubers using a hydroponic technique *Potato Research, 42*(3, 4), 457–469.

CHAPTER 13

Intelligent Smart Sensor for Cognitive Radio Networks: Comparison, Solution, and Analysis

YOGITA THAREJA, KAMAL KUMAR SHARMA, and
PARULPREET SINGH

*School of Electrical and Electronics Engineering,
Lovely Professional University, Punjab, India*

ABSTRACT

Programming configurable radio with dynamic spectrum uphold is the characteristic property of cognitive radio (CR). Interoperability of Intellectual radio with wireless sensor networks (WSNs) would empower the wireless nodes to get to and communicate the application information in authorized PU-free channels. Since wireless sensor nodes work in vigorously jam-packed ISM groups (902 MHz/2.4 GHz), there will be performance degradation in terms of spectrum lifetime and efficient spectrum sensing.

13.1 INTRODUCTION

As of late, huge advancement has been made in wireless sensor networks (WSNs). Because of this assignment in that restricted range has gotten practically incomprehensible. Along these lines, an ideal assignment in a limited spectrum is imperative to meet the recurrence interest. Since the majority of the spectrum assets are apportioned in a static way and usage

Intelligent Sensor Node-Based Systems: Applications in Engineering and Science, Anamika Ahirwar, Piyush Kumar Shukla, Prashant Kumar Shukla, and Ruby Bhatt (Eds.)

is low in a few areas, the Federal Communication Commission (FCC) expressed that the majority of the range is underutilized [1]. Intellectual radio has arisen as an effective answer for the use of the range efficiently. Intellectual radio (CR) utilizes the unused range groups in an opportunistic manner by detecting the range. The essential clients (PU) are the authorized proprietors of the range who can use the range consistently. The auxiliary clients (SU) are the unlicensed clients who can opportunistically access the channel (spatially, transiently) without making interference with the PUs. A SU ought to abandon its range band and switch to another empty range band when the PU shows up.

According to the WSN architecture, it may consist of hundreds of sensor nodes that are deployed randomly in the monitoring sensor field. The source node and base station (BS) are the two components of WSNs for data collection. The source noes transmit the data to the BS with the help of a single or multi-hop data transmission scheme.

Figure 13.1 depicts the WSN's Architecture. The all-inclusive number of remote gadgets working in industrial, scientific, and medical bands which is the ISM band makes the strict check on wireless sensors which are using this band. Besides, the expanded number of remote gadgets advances range shortage issues on account of the accessibility of restricted assets. Furthermore, framework throughput relies upon the utilization of the channel [2]. In this manner, impedance, channel use, and range of the board stay testing issues in remote correspondence.

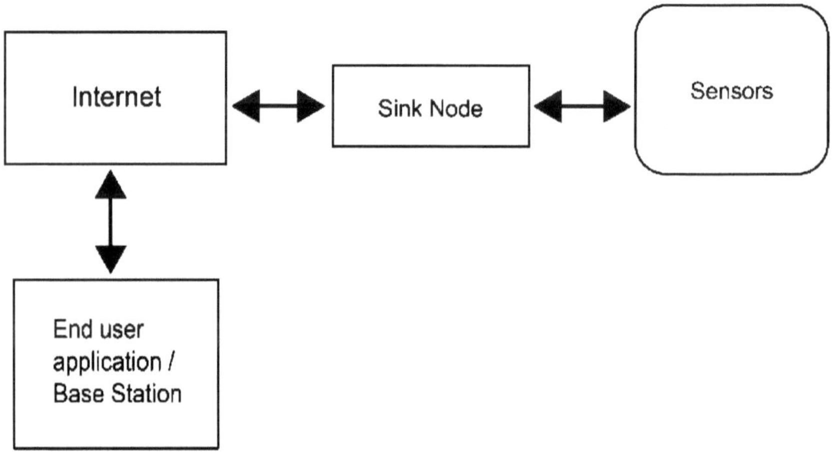

FIGURE 13.1 Architecture of WSN.

13.1.1 INTELLECTUAL RADIO: A LEAP FORWARD IN WSN

Lately, cognitive radio (CR) conspires built up a strict response to the issues of scarcity of spectrum resources by improving the utilization of the available resources. The network of CR, which is termed cognitive radio networks (CRNs) designing includes two types of users which are main users or essential (Primary) and auxiliary (Secondary) customers. The essential users (Primary users, i.e., PU) are the supported customers permitted to get to the range constantly while SUs (Secondary Users) are allowed to get to the range artfully. Whenever the range of PU is not utilized then the discretionary customers can avail of it. This improving methodology helps to reach utilization and calm the issue of blockage of the band which is not licensed [3]. In the network of CR, the auxiliary users which are also called SU are not allowed to make any impedance to the main users.

Dissimilar to the regular WSNs, the CR sensor networks work on the authorized side. Like the network of CR, main users have straight admittance towards the range while the auxiliary is permitted to get to the range entrepreneurially. Then again, CRSN embraces the overall constraints of customary WSNs like energy utilization and organization lifetime. Besides, range detecting is likewise used as an urgent angle in the execution of CR sensor networks. The effect of SS in energy utilization since a larger or greater number of SUs for detection requires more energy. Subsequently, momentum advances are centered around the effective methodology for range detecting. Also, steering and asset designation is a difficult errand in CRSNs. By exploiting the range usage rule of FCC and progressions of systems administration innovation, the consolidated remote sensor organization and CR can moderate difficult issues like range use, and asset allotment [23–39].

Figure 13.2 explains the CR which includes sensing the environment, analysis in a rapid way, reasoning in a strategic way, adaptation of the parameters which are given, and then going back again to the sensing way. The radio environment is placed in the middle, which is communicating with the sensing which is also considered as the monitoring in real-time. It is also communicating with the reasoning which is responsible for determining strategy. It is also linked with the adaptation which transition to new operating parameters. It is also connected to the analysis which analysis the environment rapidly [40–49].

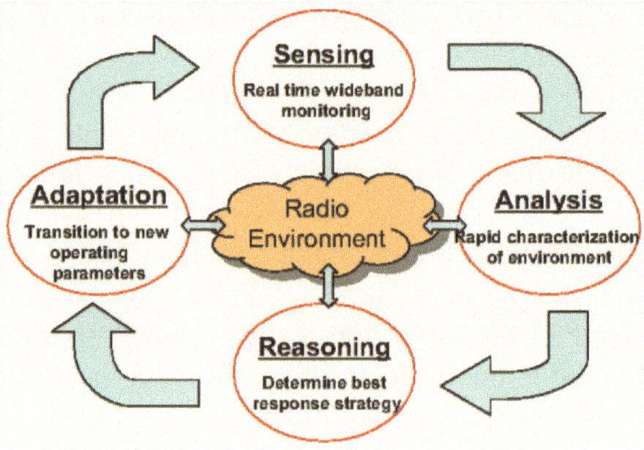

FIGURE 13.2 Cognitive cycle.

13.1.2 OVERLAY, UNDERLAY, INTERWEAVE

An intellectual radio is a remote specialized gadget that wisely uses any accessible side data about the (i) action; (ii) channel conditions; (iii) encoding techniques; or (iv) communicated information groupings of essential clients with which it shares the range.

In light of the kind of accessible organization-side data alongside the administrative imperatives, optional clients look to underlay, overlay, or interlace their signs with those of essential clients without fundamentally affecting these clients.

There are three fundamental psychological radio organization ideal models: underlay, overlay, and interlace. The underlay worldview permits optional clients to work if the obstruction, they cause to essential clients is under a given limit or meets a given bound on essential client execution debasement [50–55]. In overlay frameworks, the auxiliary clients catch the transmissions of the essential clients, then, at that point utilize this data alongside refined sign handling and coding strategies to keep up with or work on the presentation of essential clients, while likewise acquiring some extra data transfer capacity for their own correspondence. Under ideal conditions, complex encoding and translating techniques permit both the optional and essential clients to eliminate all or part of the obstruction brought about by different clients. In intertwined, frameworks the auxiliary clients distinguish the shortfall of essential client signals in space, time, or recurrence, and craftily convey during these nonattendances. For each of the three ideal models, in case there are various optional clients then, at that point, these clients should share data transfer capacity among themselves just as with the essential clients, subject to their given psychological worldview. This brings about the medium access control (Macintosh) issue among auxiliary clients, like what emerges among clients in customary remote organizations. Given this comparability, Macintosh conventions that have been proposed for optional clients inside a specific worldview are regularly determined from regular Macintosh conventions. Also, various auxiliary clients may send to a solitary auxiliary recipient, as in the uplink of a cell or satellite framework, and one auxiliary client may send to various optional beneficiaries, as in the comparing downlink. Figure 13.3 depicts CR in underlay and overlay.

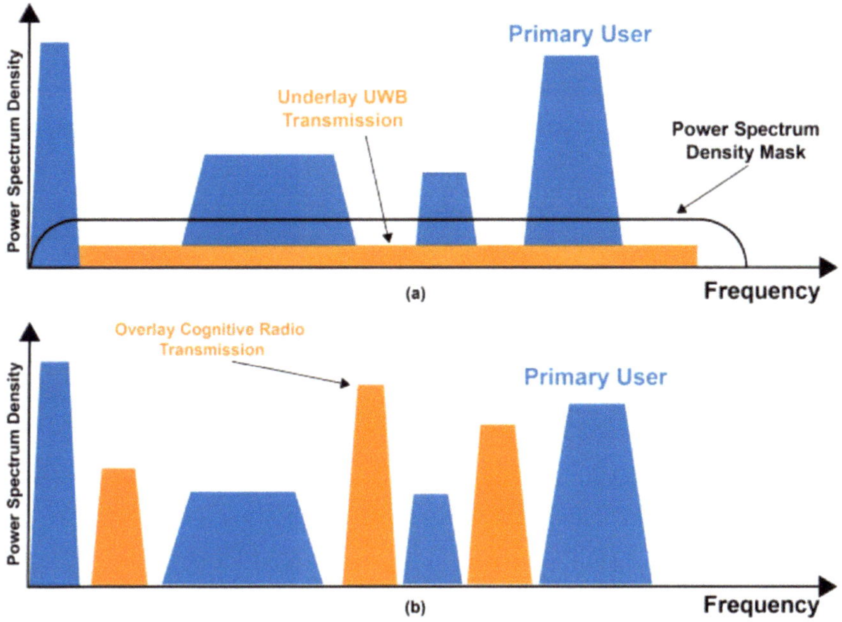

FIGURE 13.3 Cognitive radio (a) underlay and (b) overlay.

Table 13.1 depicts the difference between underlay, overlay, and interweave.

TABLE 13.1 Differentiate Among Underlay, Overlay, and Interweave

	Underlay	Interweave	Overlay
Network information side	Optional transmitters (secondary users) know obstruction caused to essential beneficiaries (primary users).	Optional clients distinguish range openings in space, time, or potential recurrence from which the essential clients are missing.	Optional hubs realize channel gains, encoding methods what's more, potentially the sent information groupings of the essential clients.
Simultaneous transmission	Optional clients (secondary users) can send at the same time as the essential clients (primary users) as long as the obstruction caused is underneath a satisfactory breaking point.	Optional clients communicate at the same time with an essential client just when there is missed recognition of the essential client action.	Optional clients can communicate all the while with the essential clients; the impedance to the essential clients can be balanced by utilizing part of the auxiliary clients' capacity to transfer the essential clients' information successions.
Power limits to transmit	Optional clients' communication power is restricted by an imperative on the impedance caused to the essential clients.	Optional clients send power is restricted by the scope of essential client action it can identify (alone or through agreeable detecting).	Auxiliary clients can send at any force, and the impedance to essential clients can be counterbalanced by handing off the essential clients' information arrangements.
Hardware	Secondary users are aware of the interference of the primary one.	It also listens to the interference of the primary users. Here the paradigms of encoding and decoding complexity are higher.	The receiver is highly agile in finding the white spaces or holes.

13.2 CONTRIBUTION OF WORK

Here the short conversation about the latest method in the field of wireless along with CR and its network is depicted. The most important one here is the sensing of the spectrum which is affecting the lifetime when the integration of WSNs is taken into account with the psychological radio. Different explores have been done for viable range the board like detecting the spectrum, dynamic range access, channel designation, and access. Besides, asset the executives and energy-mindful steering plans are likewise introduced. Afterward, the range portion issue is formed as a non-direct puzzle with the functionality of programming that can be addressed in Ref. [5]. Like this in Ref. [4], utilization of the game way by Stackelberg improves the security issues in CRSNs. Grouping is otherwise called a significant piece of steering in these organizations. Under bunching association, the path to the structure is through the group head amounts to individuals from the group who exhibit the geography of the unique non-covering groups of sensors point in the system. There are various approaches concerning gathering with changed applications in planning. Another aspect of the WSN includes the selection of the head of cluster which is notated by CH [6], which is done either by BS or not. Based on that the classification is with the centralized system or without the centralized system.

Further, the clustering protocol's division is divided as follows (Figure 13.4 and Table 13.2):

1. **Centralized Clustering:** In the clustering, which is centralized, the central BS selects the cluster heads (CH) on the basis of the size of the cluster, number of the cluster and location of CH (cluster's heads), type of the node, etc.
2. **Decentralized Clustering Protocols:** In decentralized clustering, the sink or BS will not decide the formation of the cluster. It has subsequent stages where the initial one includes the advancement of the algorithm with p states of choosing the head of the cluster.

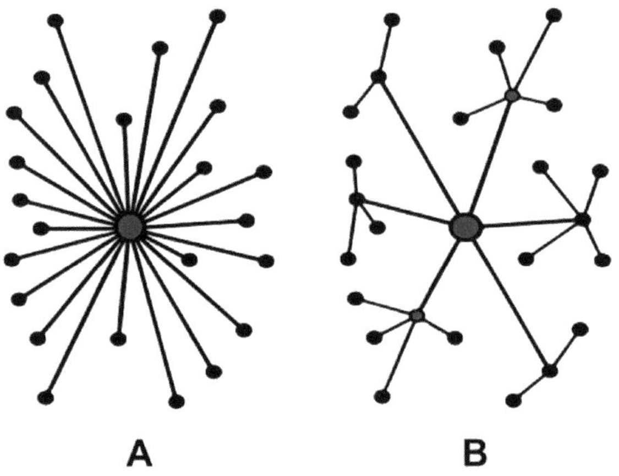

FIGURE 13.4 Clustering – (a) centralized; and (b) de-centralized.

TABLE 13.2 Merits and Demerits of Clustering Scheme

Types of Clustering	Advantages	Example
Clustering (centralized)	Cluster head dispersion is normally in the system	LEACH-C
	management is good in terms of energy	PEGASIS [10]
Clustering (decentralized)	Good network lifetime	LEACH [14]
	Communication cost is low	DEEP [11]
	Throughputs improvise	HEED [12]
	Energy consumption is less	
	Less energy consumption	EEHC [13]

The show of the protocol of bunching to the extent offering incredible results having vitality the board, parcel conveyance or all the more all extension of framework lifetime depends upon a proficient approach towards the development of reasonable bunches. Regardless, with the improvement of the frameworks that will without a doubt cover colossal geological areas, the arrangement of amazing groups will overall be hard. Thusly, the colossal geological zones and the unconstrained center transport stay against fruitful grouping frameworks, it transforms into a test when it is considered in agreeable correspondences. The different investigations are clarified in Table 13.3 with goals, benefits, and detriments.

TABLE 13.3 Analysis with Comparison of Aim, Merits, and Demerits

Aim	Drawbacks	Merits	References
Comparison of LEACH with its all versions.	• Oldest convention. • Hybrid convention required.	The head of cluster choice depends on: • Proper placement of hubs. • Proper hub distance from the base station.	[11]
Study on the wireless sensor network of psychological with hub bunching.	Proceed to CR-WSNs with: • Vitality protection. • Adjusting on loads. • PU insurance.	The decision of the group head and cluster part is principally dependent on: • Bandwidth accessibility. • Channel information.	[12]
Energy-based essential sign location system for CRSNs to break down the discovery execution.	Proceed to security.	Development QoD (quality of detection) prerequisites.	[13]

TABLE 13.3 *(Continued)*

Aim	Drawbacks	Merits	References
Range detection against attacks and the improvement of energy viability in the network of the psychological radio.	Continue different assaults.	The network of cognitive radio is helpless against range-detecting information adulteration (SSDF) assaults.	[14]
Distinguish an inexact ideal arrangement in sensible time utilizing game hypothesis and a helpful approach.	• Inefficient for an enormous sensor organization. • Faster calculation with immaterial loss of accuracy.	• Fair range task. • Avoiding futile range handoff.	[15]
• Reduction of delays. • Utilization of energy properly.	The area of the sink medium is taken to proceed further.	• Data transmission delay is reduced by about 36.6%. • Residual energy usage is required.	[16]
• Improve the stability of the networks. • Consumption of energy taken.	Proceed to interference cancellation.	• Network stability-aware clustering protocol.	[17]
• Improves transmission of data.	Test the proposed calculation in the savvy local area, web of vehicles, and biological observing.	• A clustering algorithm with spectrum awareness.	[18]
• SNS clustering. • Castalia simulator is taken for outputs.	Proceed to inter-cluster interference.	• Network lifetime is improvised. • The itinerary length is reduced.	[20]
• Throughput is enhanced. • Delays are reduced. • AODV protocol is minimized energy consumption.	Heterogeneous wireless networks are taken into account.	• Power controlling algorithm. • Reduced unwanted channel switching.	[21]

13.3 DISCUSSION AND RESULTS

Here we explained the total exhaustive test study utilizing the proposed spectrum-detecting model and energy-aware clustering for CRSNs in MATLAB (Table 13.4).

TABLE 13.4 Parameters (Simulation)

Parameter Name	Parameter Value
CW minimum size for RT traffic	8
Number of nodes	2–10
CW minimum size for NRT traffic	32
Population size	100
Maximum generation	250
Sampling times TO	5
Transmission rate	250 kb/s

We present the relative examination as far as range detecting, and phony alert for different psychological wireless hubs. The result has been received from Ref. [22] where multiple destinations are taken for the network of CR. Then we explained the detection of the spectrum with 2 to 10 sensor hubs taken into account. The relative comparison explained that our proposed technique is better.

In the same manner, we measured the fake alarm's performance and compared the result with state-of-art techniques (Figure 13.5).

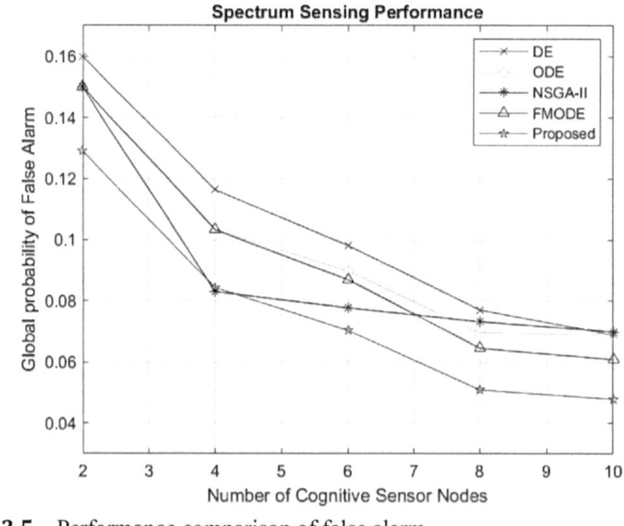

FIGURE 13.5 Performance comparison of false alarm.

In the networks of CR, the usage of channels plays an urgent role and has a lot of impact on the performance of the networks. Here we have considered the usage parameter of the channel and compared the performance of the usage in Ref. [22]. The techniques which are explained in Ref. [22] such as iPCMAC and PCMAC protocol are based on the integration of CRs with medium access protocol whereas the proposed one which is here focuses on resource and energy-aware routing (Figure 13.6).

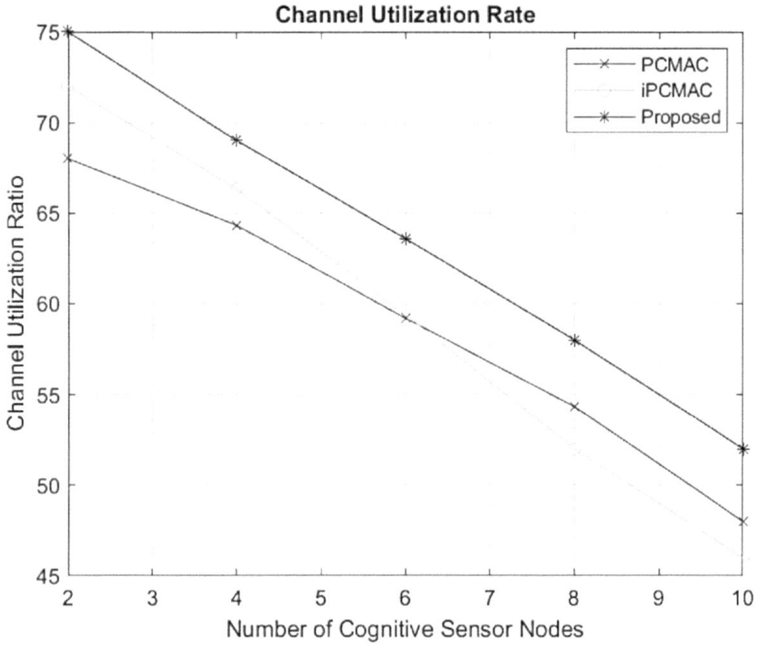

FIGURE 13.6 Usage rate (channel).

The relative experimentation shows that recommended technique causes improvement of channel usage rate which helps to reduce asset wastages and upgrade the organization execution.

13.4 CONCLUSION

In this exploration examines a novel method to manage and consolidate the Sensors Nodes with Cognitive hubs to course sensor data to sink using approved innovatively. Diverse grouping methods are clarified above for certain benefits and impediments. In the above segment, bunching plans were named incorporated, and decentralized with appropriate models. The challenges being utilized of bunching conventions recalled inconveniences for making groups. This is where the flow of hubs in the framework is unbalanced and even more so exactly when the framework is covering a broader zone. The show on the use of bunches completed ideas of future work that included staggered grouping, versatile Quality of administration bunching, and steering measurements that connect holes clarified. The improvement in the participation of clients includes agreeable detecting. The usage of the Channel proportion should be expanded for certain half-and-half conventions without affecting QoS. The bogus caution rate is likewise needed to improve further.

ACKNOWLEDGMENT

I'm profoundly appreciative to Vivekananda Institute of Professional Studies, Delhi, India for providing me framework and Labs to finish my work.

KEYWORDS

- **cluster communication radio**
- **cognitive radio**
- **cognitive radio networks**
- **cognitive radio sensor networks**
- **network lifetime**
- **spectrum sensing**

REFERENCES

1. Salameh, H. A. B., & Krunz, M., (2009). Channel access protocols for multihop opportunistic networks: Challenges and recent developments. *IEEE Netw., 23*, 14–19.

2. So, J., & Srikant, R., (2015). Improving channel utilization via cooperative spectrum sensing with opportunistic feedback in cognitive radio networks. *IEEE Communications Letters, 19*(6), 1065–1068.
3. Wyglinski, A. M., et al., (2009). *Cognitive Radio Communications and Networks: Principles and Practice*. Academic Press.
4. Abdalzaher, M. S., Seddik, K., & Muta, O., (2017). Using Stackelberg game to enhance cognitive radio sensor networks security. *IET Communications, 11*(9), 1503–1511.
5. Zeng, B., Zhang, C., Hu, P., & Wang, S., (2017). Spectrum sharing based on a Bertrand game in cognitive radio sensor networks. *Sensors, 17*(1), 101.
6. Sucasas, V., et al., (2016). A survey on clustering techniques for cooperative wireless networks. *Ad. Hoc. Networks, 47*, 53–81.
7. Lindsey, S., & Raghavendra, C. S., (2002). PEGASIS: Power-efficient gathering in sensor information systems. In: *Proceedings, IEEE Aerospace Conference* (Vol. 3, pp. 3–5). IEEE.
8. Veyseh, M., Wei, B., & Mir, N. F., (2006). An energy-efficient decentralized clustering protocol for wireless sensor networks. In: *Proc. 5 WSEAS Int. Conf. Telecommun. Informat.* (pp. 293–298).
9. Xu, J., Jin, et al., (2012). Improvement of LEACH protocol for WSN. In: *2012 9 International Conference on Fuzzy Systems and Knowledge Discovery* (pp. 2174–2177). IEEE.
10. Kumar, D., Aseri, T. C., & Patel, R. B., (2009). EEHC: Energy efficient heterogeneous clustered scheme for wireless sensor networks. *Computer Communications. 32*(4), 662–667.
11. Dabas, P., & Gupta, N., (2015). LEACH and its improved versions-a survey. *International J. Sci. Eng. Res., 6*, 184–188.
12. Joshi, G. P., & Kim, S. W., (2016). A survey on node clustering in cognitive radio wireless sensor networks. *Sensors, 16*, 1465.
13. Tang, S., Trautman, A., & Nguyen, H., (2016). On energy-based quality of detection (QoD) for cognitive radio sensor networks. *International Journal of Wireless Information Networks, 23*(3), 214–221.
14. Pawan, W., & Ahmad, S. N., (2014). Simulation of linear dynamic macro model of photocatalytic sensor in SPICE. *Compel, the International Journal of Computation and Mathematics in Electrical and Electronic Engineering, 33*(1, 2). ISSN: 0332-1649 (SCI, ISI index).
15. Byun, S. S., et al., (2017). Fair dynamic spectrum allocation using modified game theory for resource-constrained cognitive wireless sensor networks. *Symmetry, 9*(5), 73.
16. Byu, S. S., & Gil, J. M., (2017). Fair dynamic spectrum allocation using modified game theory for resource-constrained cognitive wireless sensor networks. *Symmetry, 9*(5), 73.
17. Liu, A., Chen, W., & Liu, X., (2018). Delay optimal opportunistic pipeline routing scheme for cognitive radio sensor networks. *International Journal of Distributed Sensor Networks, 14*(4), 1550147718772532.

18. Zheng, M., Chen, S., Liang, W., & Song, M., (2019). NSAC: A novel clustering protocol in cognitive radio sensor networks for internet of things. *IEEE Internet of Things Journal.*
19. Idoudi, H., Mabrouk, O., Minet, P., & Saidane, L. A., (2019). Cluster-based scheduling for cognitive radio sensor networks. *Journal of Ambient Intelligence and Humanized Computing, 10*(2), 477–489.
20. Wang, T., Guan, X., Wan, X., Shen, H., & Zhu, X., (2019). A spectrum-aware clustering algorithm based on weighted clustering metric in cognitive radio sensor networks. *IEEE Access.*
21. Shah, S. B., Chen, Z., Yin, F., Khan, I. U., & Ahmad, N., (2018). Energy and interoperable aware routing for throughput optimization in clustered IoT-wireless sensor networks. *Future Generation Computer Systems, 81*, 372–381.
22. El Fissaoui, M., Beni-Hssane, A., & Saadi, M., (2019). Energy-efficient and fault tolerant distributed algorithm for data aggregation in wireless sensor networks. *Journal of Ambient Intelligence and Humanized Computing, 10*(2), 569–578.
23. Carie, A., Li, M., Marapelli, B., Reddy, P., Dino, H., & Gohar, M., (2019). Cognitive radio assisted WSN with interference aware AODV routing protocol. *Journal of Ambient Intelligence and Humanized Computing, 10*(10), 4033–4042.
24. Raymond, J. W., Olwal, T. O., & Kurien, A. M., (2018). Cooperative communications in machine to machine (M2M): Solutions, challenges and future work. *IEEE Access, 6*, 9750–9976.
25. Al-Medhwahi, M., Hashim, F., Ali, B. M., Sali, A., & Alkholidi, A., (2019). Resource allocation in heterogeneous cognitive radio sensor networks. *International Journal of Distributed Sensor Networks, 15*(7), 1550147719851944.
26. Bayrakdar, M. E., (2020). Cooperative communication-based access technique for sensor networks. *International Journal of Electronics, 107*(2), 212–225.
27. Cicioğlu, M., Bayrakdar, M. E., & Çalhan, A., (2019). Performance analysis of a new MAC protocol for wireless cognitive radio networks. *Wireless Personal Communications, 108*(1), 67–86.
28. Bayrakdar, M. E., & Çalhan, A., (2017). Performance analysis of sensing-based spectrum handoff process for channel bonding mechanism in wireless cognitive networks. In: *2017 25 Signal Processing and Communications Applications Conference (SIU)* (pp. 1–4). IEEE.
29. Ali, A., Abbas, et al., (2019). Hybrid fuzzy logic scheme for efficient channel utilization in cognitive radio networks. *IEEE Access, 7*, 24463–24476.
30. Ali, A., et al., (2018). RaptorQ-based efficient multimedia transmission over cooperative cellular cognitive radio networks. *IEEE Trans. Veh. Technol., 67*(8), 7275–7289.
31. Wu, C., Wang, Y., & Yin, Z., (2018). Energy-efficiency opportunistic spectrum allocation in cognitive wireless sensor network. *EURASIP Journal on Wireless Communications and Networking, 2018*(1), 13.
32. Doung, A., (2012). *Cognitive Radio: Advantages and Challenges*. PH250, Stanford University, Spring.
33. Rahul, R. N., et al., (2020). Novel economical social distancing smart device for COVID-19. *International Journal of Electrical Engineering and Technology, 11*(4), 204–217.

34. Nadikattu, R. R., (2020). *Information Technologies: Rebooting the World Activities During COVID-19.* https://www.kije.com.ng/index.php/KIJE/article/view/18 (accessed on 1 March 2023).
35. Liu, W., Qin, G., Li, S., He, J., & Zhang, X., (2015). A multiobjective evolutionary algorithm for energy-efficient cooperative spectrum sensing in cognitive radio sensor network. *International Journal of Distributed Sensor Networks, 11*(5), 581589.
36. Vaibhav, B., & Pawan, W., (2013). A secured dual tone multi-frequency based smart elevator control system. *International Journal of Research in Engineering and Advanced Technology. 4*(1). ISSN (Online): 2319-1163.
37. Pawan, W., & Ahmad, S. N., (2013). A novel pseudo NMOS integrated ISFET device for water quality monitoring. *Active and Passive Components* (Vol. 1, No. 1). Hindawi Article ID: 258970, (Scopus). ISSN: 0882-7516.
38. Vaibhav, B., & Pawan, W., (2013). Modeling and simulation of electrical load control system using RF technology. *International Journal of Multidisciplinary Science and Engineering, 4*(2), 44–47. ISSN 2045-7057.
39. Pawan, W., & Ahmad, S. N., (2014). Development of economical ASIC For PCS for water quality monitoring. *Journal of Circuit System and Computers, 23*(6), 1–13. ISSN: 0218-1266 (SCI, ISI index).
40. Shaikh, F. K., & Zeadally, S., (2016). Energy harvesting in wireless sensor networks: A comprehensive review. *Renewable and Sustainable Energy Reviews, 55*, 1041–1054.
41. Raza, M., Aslam, N., Le-Minh, H., Hussain, S., Cao, Y., & Khan, N. M., (2017). A critical analysis of research potential, challenges, and future directives in industrial wireless sensor networks. *IEEE Communications Surveys & Tutorials, 20*(1), 39–95.
42. Görgü, L., Kroon, B., O'Grady, M. J., Yılmaz, Ö., & O'Hare, G. M., (2018). Sensor discovery in ambient IoT ecosystems. *J. Ambient Intell. Hum. Comput., 9*, 447–458.
43. Zhang, F., Jing, T., Huo, Y., & Jiang, K., (2019). Joint optimization of spectrum sensing and transmit power in energy harvesting cognitive radio sensor networks. *The Computer Journal, 62*(2), 215–230.
44. Akan, O. B., Karli, O. B., & Ergul, O., (2009). Cognitive radio sensor networks. *IEEE Network, 23*(4), 34–40.
45. Ali, A., & Hamouda, W., (2016). Advances on spectrum sensing for cognitive radio networks: Theory and applications. *IEEE Communications Surveys & Tutorials, 19*(2), 1277–1304.
46. Darsena, D., Gelli, G., & Verde, F., (2017). An opportunistic spectrum access scheme for multicarrier cognitive sensor networks. *IEEE Sensors Journal, 17*(8), 2596–2606.
47. Mohjazi, L., Dianati, M., Karagiannidis, G. K., Muhaidat, S., & AlQutayri, M., (2015). RF-powered cognitive radio networks: Technical challenges and limitations. *IEEE Communications Magazine, 53*, 94–100.
48. Mikeka, C., & Arai, H., (2011). Design issues in radio frequency energy harvesting system. In: *Sustainable Energy Harvesting Technologies Past, Present, and Future* (pp. 235–256). InTech.
49. Sudevalayam, S., & Kulkarni, P., (2010). Energy harvesting sensor nodes: Survey and implications. *IEEE Communications Surveys & Tutorials, 13*, 443–461.
50. Lu, X., Wang, P., Niyato, D., Kim, D. I., & Han, Z., (2014). Wireless networks with RF energy harvesting: A contemporary survey. *IEEE Communications Surveys & Tutorials, 1*, 757–789.

51. Singla, J., Mahajan, R., & Bagai, D., (2018). A survey on energy harvesting cognitive radio networks. In: *2018 6 Edition of International Conference on Wireless Networks & Embedded Systems (WECON)* (pp. 6–10). IEEE.
52. Bhowmick, Roy, s., & Kundu, S., (2016). Throughput of a cognitive radio network with energy-harvesting based on primary user signal. *IEEE Wireless Commun. Lett., 5*(2), 136–139.
53. Gao, Y., He, H., Deng, Z., & Zhang, X., (2018). Cognitive radio network with energy harvesting based on primary and secondary user signals. *IEEE Access, 6*, 9081–9090.
54. Shafie, A. E., Ashour, M., Khattab, T., & Mohamed, A., (2015). On spectrum sharing between energy harvesting cognitive radio users and primary users. In: *Proc. Int. Conf. Comput., Netw. Commun.* (pp. 214–220). Garden Grove, CA, USA.
55. Fadel, E., Faheem, M., Gungor, V. C., Nassef, L., Akkari, N., Malik, M. G. A., & Akyildiz, I. F., (2017). Spectrum-aware bio-inspired routing in cognitive radio sensor networks for smart grid applications. *Computer Communications, 101*, 106–120.

CHAPTER 14

Smart and Ecofriendly Intelligent House Based on IoT and Simulation Using a Cisco Networking Simulator

RAVI RAY CHAUDHARI, KRISHNA KUMAR JOSHI, NEELAM JOSHI, and ANAND KUMAR PANDEY

Department of Computer Engineering, SKNSITS Lonavala, India

Department of Computer Engineering, SITS Lonavala, India

Department of Computer Science and Engineering, ITM University Gwalior, India

ABSTRACT

Currently, the whole world is witnessing an era of technology. At this point, everyone is fully set up on digital devices. You can use the digital devices to protect your home and at the same time, you can look and manage your home through online (e.g., fan, air conditioner, television, home window, door, shutter, cooling, etc.). So, in this chapter, we will understand how we can control everything (electrical) in your home, and how we can adapt to the home environment. We have compiled this whole chapter with Cisco Packet Tracer Networking Simulator to help you and protect your home and also you can watch your home from anywhere. With the help of the IoT concept, you can manage and monitor all devices running on all batteries and the flowing machines running with the heat of the wire. You can also manage and monitor them.

Intelligent Sensor Node-Based Systems: Applications in Engineering and Science, Anamika Ahirwar, Piyush Kumar Shukla, Prashant Kumar Shukla, and Ruby Bhatt (Eds.)

At this time, thanks to all the people who are very busy and working on this project, we can protect our house. So, with the help of this chapter here, we can turn off everything electrical in our house. To perform this function, we must use a security method, in which case we must provide the password center and the emergency center with care.

Most of the time, we see that most homes are smart these days; we use routers connecting devices with the help of all the electrical devices shown and managed [1, 2]. There we built a wireless network, and we can use it to control our device electronically with a wireless signal. But with the advent of technology and this version of Cisco Packet Tracer 7.20, we can monitor and manage all the equipment in our home.

14.1 INTRODUCTION

With the help of Smart Technology Internet of Things, we can monitor all household activities efficiently. In a way, we can say that we can protect our homes without interference. It can also transform the stratosphere of a house into a stratosphere. With the help of this technology, we can take care of the safety of our homes even if we live elsewhere [3, 4]. With this technology, we can make the house airtight and safe by using sensors. Why do we use the latest Internet technology here, when we take care of our house anywhere, which is very important in modern times? Where the internet of things (IoT) is the latest technology. All of these technologies are electronic devices and psychological technologies that demonstrate the transfer of transmission between sensory-based devices. Here we complete this task by building a network created using the Cisco Packet Tracer 7.20 networking simulator [6, 7]. And by using this simulator (Cisco Packet Tracer 7.20), we will show that anyone can monitor each of the technology items in their home using the internet of things (IoT). The use and design of all these technologies are illustrated in this chapter.

14.1.1 MOTIVATION

Cisco Packet Tracer is an excellent simulator for building or simulating a network, we can understand how your physical activity is used with a virtual medium. Where Cisco Packet Tracer describes the process of connecting multiple communication devices and data communications between them. And Cisco can build a new network with the help of a packet tracer, giving us more communication equipment, front network, and wireless, we have IOE methods such as SBC-PT and MCU-PT also strengthen control/random boards. By using Cisco Packet Tracer, we can store our data in a cloud-based database.

And we can run our home and office monitoring and devices using the internet and local environment, we can do all these things with the Cisco Packet Networking Simulator, all of them prepared in python and java language.

14.1.2 METHODOLOGY

If we, in modern times, all go to smart and digital technology. A detailed answer is found in this chapter on all these issues here.

Where Cisco can solve this problem with the help of a packet tracer, how to send data from one place to another. When different types of cable connections are provided in the Cisco packet tracer and many types of communication equipment are provided.

We can create a network by connecting and changing signals and data. To illustrate the point of this chapter, the Cisco Packet Tracer version 7.20 is used.

14.2 CISCO PACKET TRACER

If, according to the result below, with the help of the Cisco simulator, it can be seen that we can protect our home from any location with wireless network equipment provided under the new Cisco Packet Tracer Off that can use all electrical devices to launch inside a smart home.

And they can monitor continuously. By using Cisco Simulator, they can use the entire network. In the Cisco simulator, various types of cable connections like copper cable to axial cable, automatic cable, etc. They are offered in many types of connecting machines.

And if we look at communication devices on it, then the Cisco Packet Tracer has provided many types of switching. When the button helps send the message signal with the appropriate data and route.

It also helps to filter the data signal. Cisco simulator helps to send data efficiently on computers, laptops, Android mobile devices, and wireless communication devices.

In this chapter, we are told that sensors, smart devices, and small controls are also used. The diagram given in Figure 14.1 describes the simulation of Cisco.

With this software, we can also control and monitor electronic devices with a human voice. To emulate all of this, Java and Python can be used with the help of language and can be emulated to send data signals. The sensor contains an alarm, which will give us information about the current state of our equipment. You are in a certain situation, but you can also monitor your device.

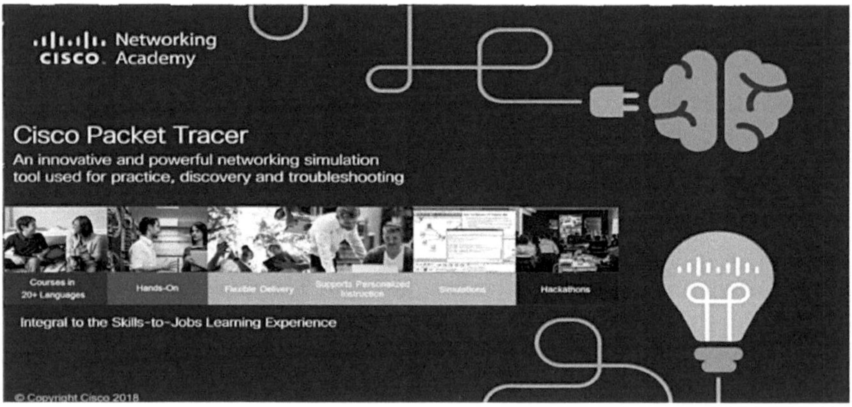

FIGURE 14.1 Cisco networking simulator.

14.3 RESULTS AND DISCUSSION

14.3.1 EFFECT OF OPERATIONAL CONDITIONAL

In Cisco packet, tracer networking simulation, IoT, router, switch, sound system, wireless networking equipment, alarm webcam switching system, etc., are all provided in it.

All of these devices can be managed via the IoT. In this chapter, it is stated that a user can monitor and use his or her home according to the internet center, without having to store photos taken on a web camera via a cloud system (Figure14.2).

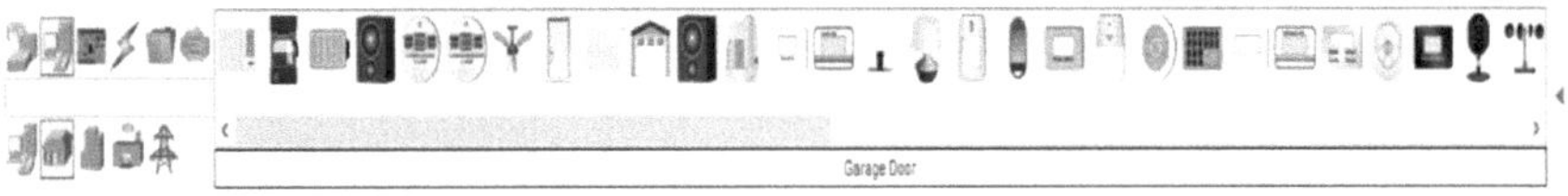

FIGURE 14.2 IoT-based components.

14.3.1.1 WIRELESS ROUTER

One of the most important communication tools in this chapter is to design a secure network to help balance remote control signals to transmit data to all electronic devices.

When everyone is working with IoT, and these users can monitor and run their homes with an Internet browser. In this whole research chapter, it has been suggested that a wireless router be used. We have four different types of Ethernet ports with "automatic" SSID compatibility to protect the establishment of WPA2 on wireless connections (Figures 14.3–14.7).

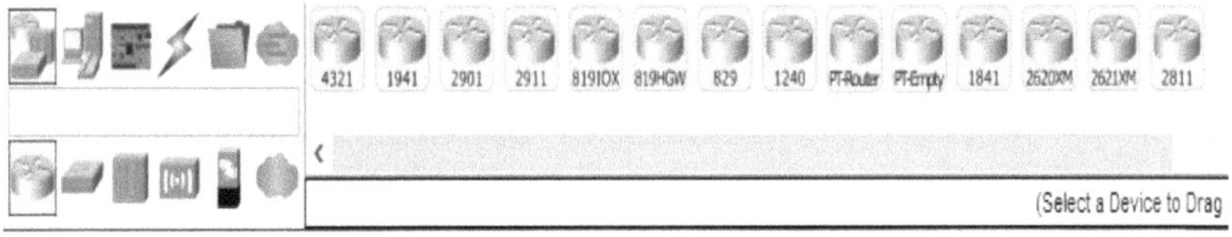

FIGURE 14.3 Different types of routers in the cisco network simulator.

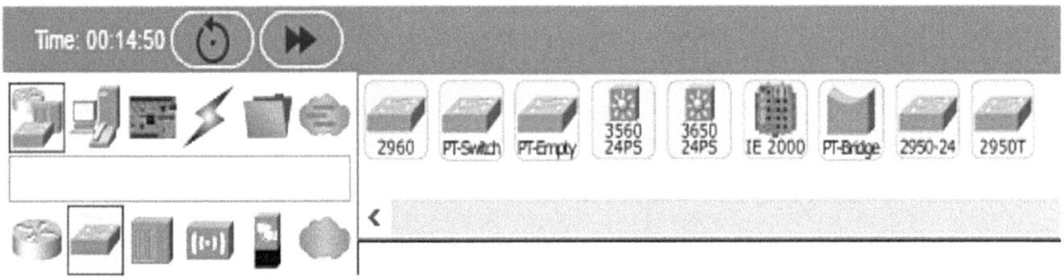

FIGURE 14.4 Different types of switches in the cisco network simulator.

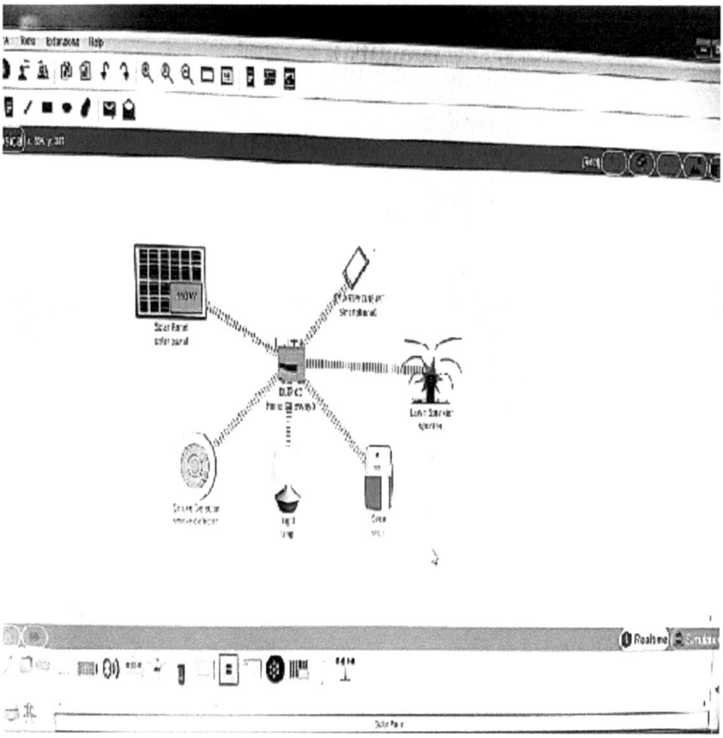

FIGURE 14.5 Subcomponent of IoT-based sprinkler.

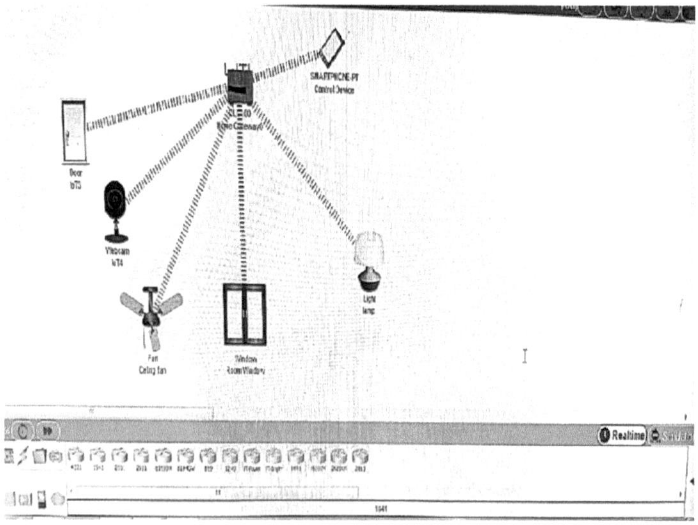

FIGURE 14.6 Pre-simulation of eco-friendly smart room.

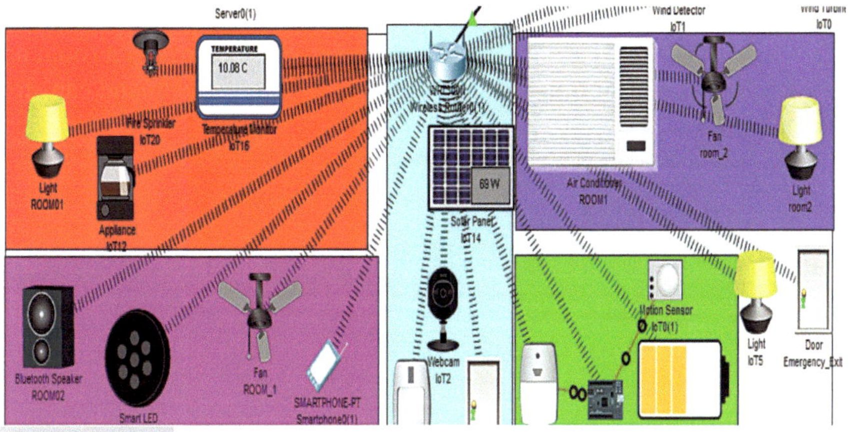

FIGURE 14.7 Wireless network router connected with the devices.

The wireless router is taken off because the user cannot transfer data via cable-based remote control. Users can use their homes from anywhere, where a wireless router can transfer data to all electronic home devices using the Internet protocol. Throughout this process, using the dynamic host configuration protocol creates automatic internet protocols so that the user can create internet protocols over and over again, so it is very important to have the dynamic host configuration protocol.

14.3.1.2 IMPLEMENTATION

By using the Cisco Packet Tracer Networking Simulator, we can build a beautiful and secure home in it. In this chapter, tell me how to make our home smarter and more secure; for this, I have used many smart devices. How to monitor and manage your home from anywhere in the world using the internet of think concept, dynamic host configuration protocol server serves as the most important character in the whole process because there is a powerful host configuration protocol server provides a temporary internet protocol address, allowing all communication components to communicate signals and messages. In this chapter, I have shown how to protect a wireless router, here is a secure pass message using a message digest to secure the entire network properly and get more light from the third user to do this, solar panels are used for this network. It is also said that when a user passes through a room and near a house, the motion detector can detect its movement and can store it in the cloud.

Simulation code in Python language:

```
fromgpio import *
from time import *
def main ():
while true:
ifdigitalRead (0) == HIGH:
print ("one more")
customWrite (1, HIGH)
sleep (100)
other:
print ("good")
custom Write (1, LOW)
sleep (5)
if __name__ == "__main__":
primary ()
```

14.3.1.3 SAFETY

This wireless router is protected with WPA2-PSK. As you can see number 5. In that, there was one large door and one emergency exit door. All IoT drawings are encrypted and registered at the server address 192.168.1.1. The username and password of all Internet object contracts are the same except for two emergency exits and IoT 8 lights. These two IoT contracts are registered with different usernames and passwords. I have done this to strengthen security. This will work for any emergency. Storing CCTV data using the cloud. Anything we can accomplish anytime and anywhere (Figures 14.8–14.17).

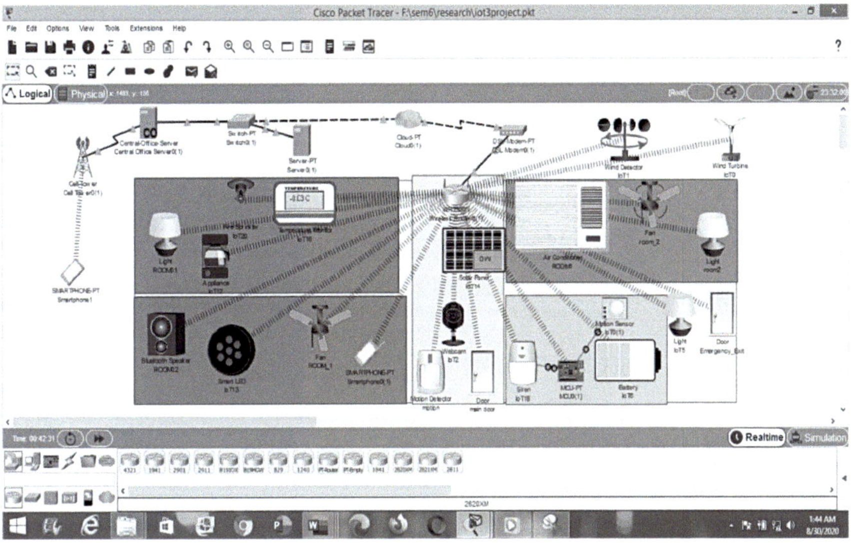

FIGURE 14.8 Design to smart and eco-friendly architecture.

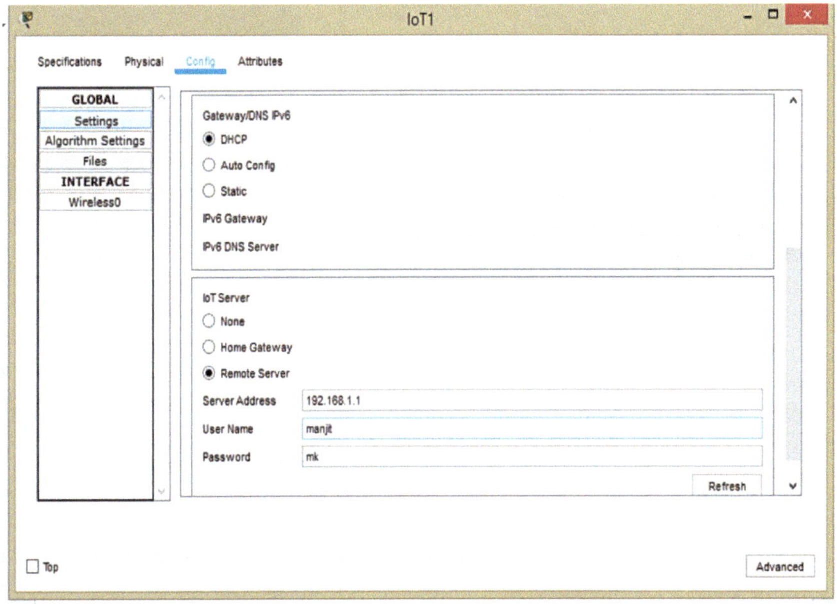

FIGURE 14.9 Setup of IoT devices in cisco network simulator.

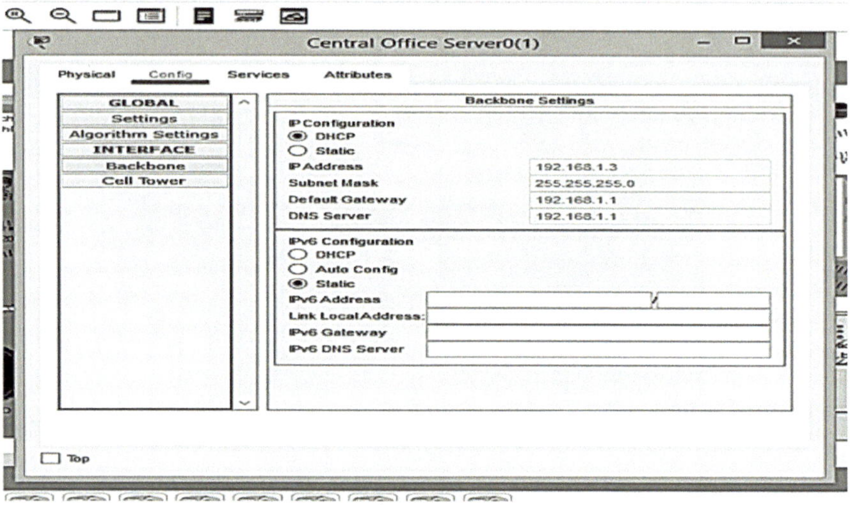

FIGURE 14.10 Main central office server.

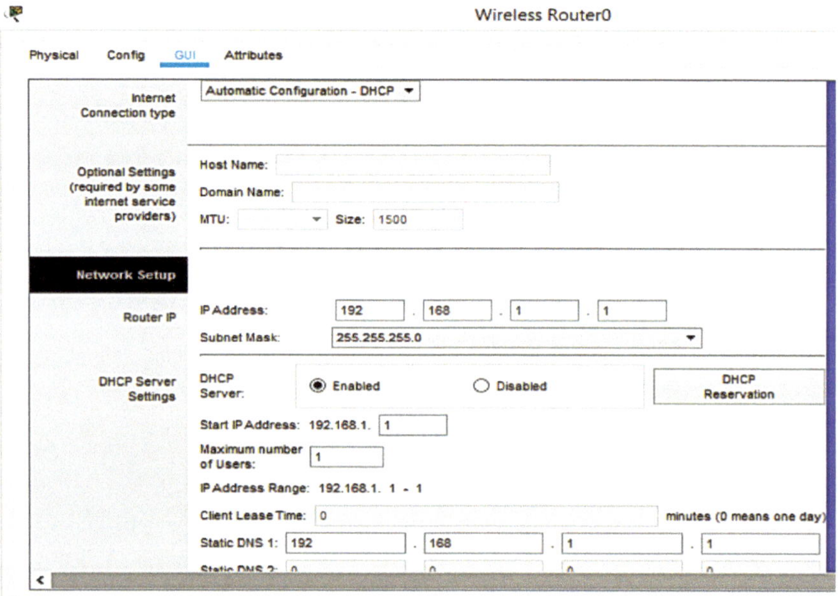

FIGURE 14.11 Configuration of wireless router setup.

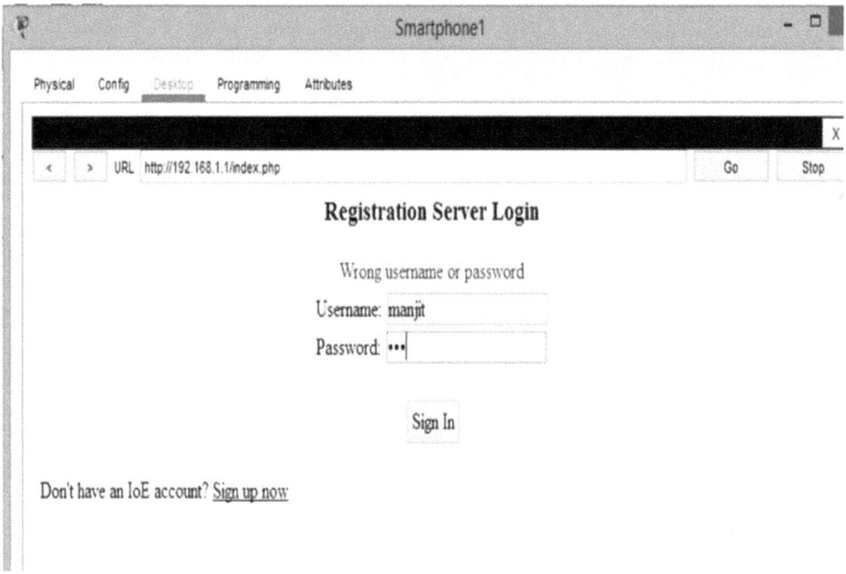

FIGURE 14.12 Set to registration server login/signup page.

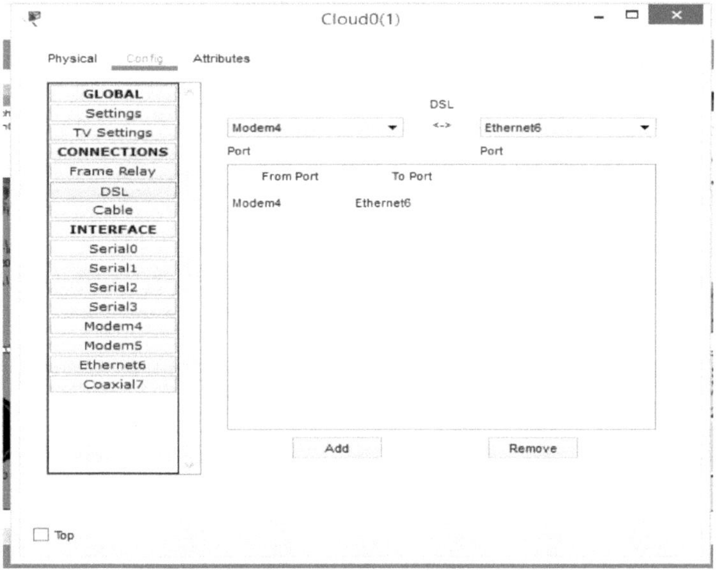

FIGURE 14.13 Design to cloud setup.

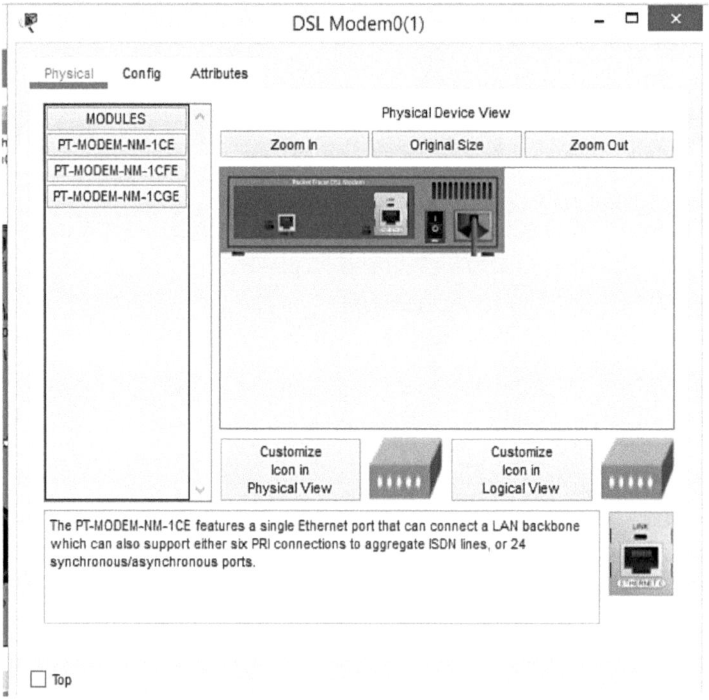

FIGURE 14.14 DSL modem.

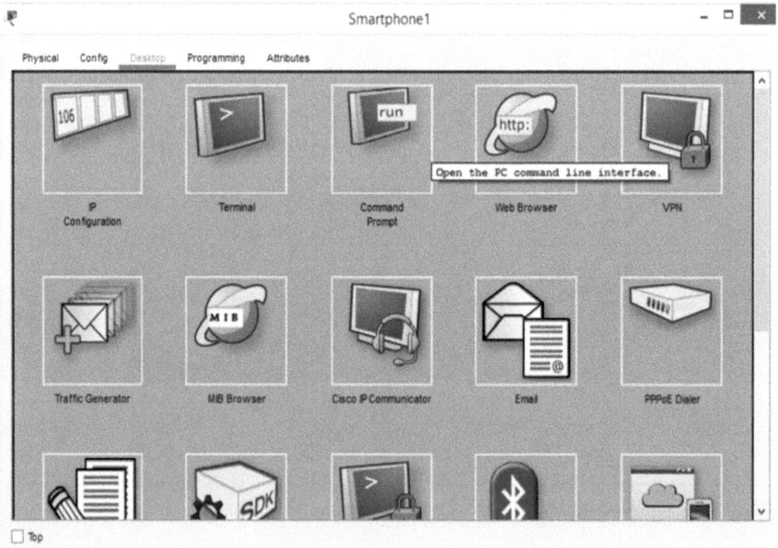

FIGURE 14.15 Smart device interface for all subcomponents.

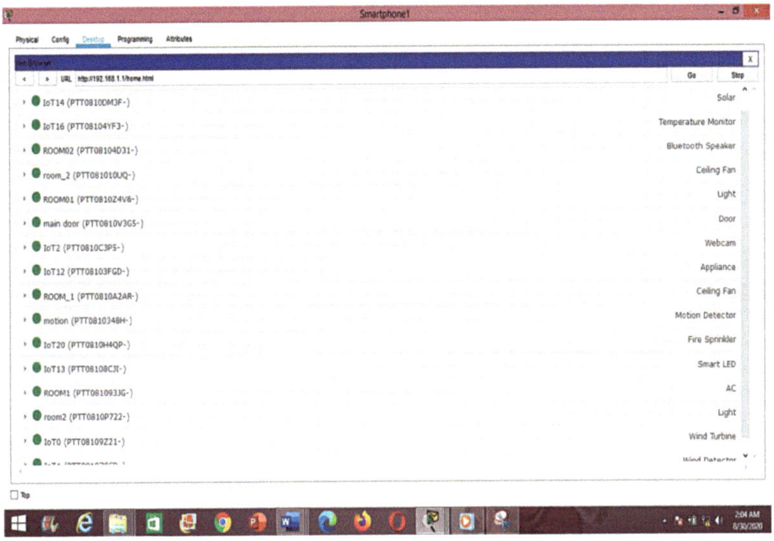

FIGURE 14.16 Registration with IoT devices.

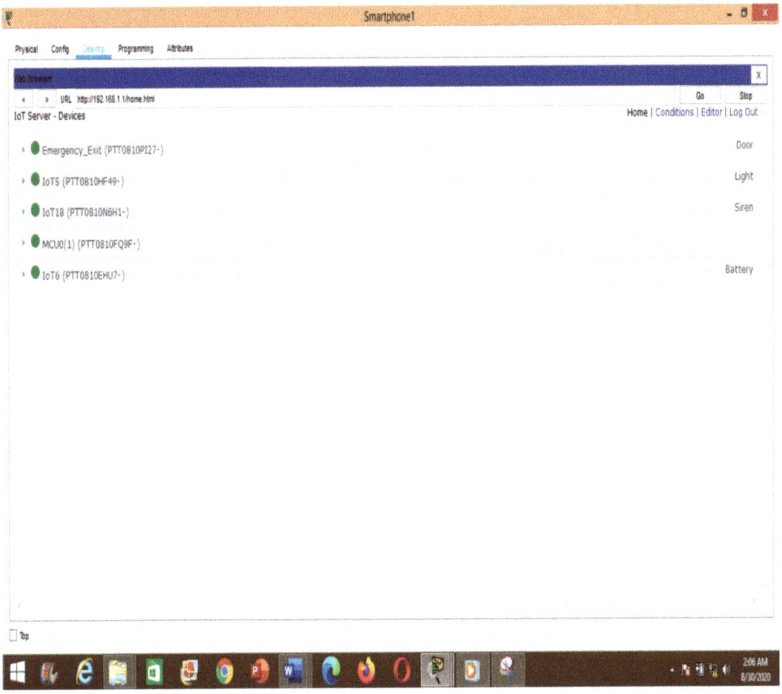

FIGURE 14.17 The working of emergency IoT devices.

14.3.1.4 CONCLUSION AND FUTURE SCOPE

In this chapter, we tried to understand how we can monitor and control all electronic devices indoors, and with the imitation of a Cisco packet tracer, we can protect the house and use the environment. With the help of internet items, you can use all gadgets with the assistance of a webcam. You can seize all of the house and home Bahar Bali plow, in this net of ideas, domestic safety passwords, and some safety codes of the solar to move all indoor gadgets. What a sun plant can be wished.

KEYWORDS

- **board**
- **cisco packet tracer**
- **home gateway**
- **Internet of Things**
- **sensors**
- **smart cell**
- **wireless router**

REFERENCES

1. Shrobe, H., Shrier, D. L., & Pentland, A., (2018). Chapter 13 – data security and privacy in the age of IoT. *New Solutions for Cybersecurity*.
2. Sharma, U., & Reddy, S. R. N., (2012). Design of home/office automation using wireless sensor network. *International Journal of Computer Applications, 43*, 53–60.
3. Haller, S., Karnouskos, S., & Schroth, C., (2009). The internet of things in an enterprise context. In: *Future Internet-FIS International Journal of Engineering Science Invention Research & Development*: *Lecture Notes in Computer Science* (Vol. 5468, No. 7, pp. 14–28).
4. Fouchal, B. H., Zytoune, O., & Aboutajdine, D., (2018). *Drip Irrigation System using Wireless Sensor Networks, 44*(4).
5. Gutiérrez, J., Villa-Medina, J. F., et al., (2014). *Automated Irrigation System Using a Wireless Sensor Network and GPRS Module, 63*(1), 166–176. IEEE.
6. Vanaja, K. J., Suresh, A., Srilatha, S., Kumar, K. V., & Bharath, M., (2018). IoT-based agriculture system using NodeMCU. *IRJET, 5*(3).

7. Gutiérrez, J., Francisco, J., Villa-Medina et al. (2014). Automated irrigation system using a wireless sensor network and GPRS module, *IEEE 63*(1), 166–176.
8. Chikankar, P. B., Mehetre, D., & Das, S., (2017). *An Automatic Irrigation System Using ZigBee in Wireless Sensor Network*, *5*(4).
9. *International Journal of Advanced Research in Computer and Communication Engineering*, *5*(6), https://ijarcce.com/volume-5-issue-6/ (accessed on 1 March 2023).

CHAPTER 15

Different Techniques of Data Fusion in the Internet of Things (IoT)

HARSH PRATAP SINGH, BHASKAR SINGH, and RASHMI SINGH

Department of Computer Science and Engineering, Sri Satya Sai University of Technology and Medical Science, Sehore, Madhya Pradesh, India

Editor in Chief, Bhopal Hundred News24, Bhopal, Madhya Pradesh, India

MIS Head, Trident Group, Budhni, Hoshangabad, Madhya Pradesh, India

ABSTRACT

The Internet of Things (IoT) is an active area of research as it has been increasing remarkably over the last few years. The IoT is a huge network of connected things and people that collect and share data about how they are used and the environment around them. It is the perception of connecting any device to the Internet and other connected devices. When something is connected to the Internet, it can send or receive information, or both. This ability to send and/or receive information makes things smart. IoT provides businesses and people a better insight into and control over the objects and environments that remain beyond the reach of the Internet. And by doing so, IoT allows businesses and people to be more connected to the world around them and to do more meaningful, higher-level work. Data fusion techniques are used to extract meaningful information from heterogeneous IoT data. It fuses distinct data from sensor sources to communally find a consequence, which is more consistent, precise, and comprehensive. This paper briefly describes the IoT in data acquisition and fusion characteristics.

Intelligent Sensor Node-Based Systems: Applications in Engineering and Science, Anamika Ahirwar, Piyush Kumar Shukla, Prashant Kumar Shukla, and Ruby Bhatt (Eds.)

15.1 INTRODUCTION

The internet of things (IoT) has almost exclusively prolonged the internet to consist of a wide range of novel features, processes, and settings. IoT makes dumb items smart by giving them a location to direct data via the Internet, permitting them to communicate with other IoT-enabled tools and individuals. IoT associates portable and digital gadgets with sensory devices connected to a network of online objects, which integrates data from innumerable devices and incorporates exceedingly respected data exchange with built-in applications. The IoT is not only limited to households. It can be used in various filed, such as devices, industries, etc. IoT Control platforms can perceive what information is vivacious and what may be overlooked with specific precision. This data can be cast-off to plug patterns, make recommendations, and envisage promising problems before they ensue. Continuous analytics offers you the power to make processes more proficient since it provides you insight. You can mechanize certified jobs using systems and smart devices, particularly if they are deadly, monotonous, time-consuming, or injurious. Seeing an IoT vision can be cast off as an example of real life. If we need to know which extra features are the utmost popular, we can use an IoT device to sync data automatically. Smart judgments are made by associated devices based on real-time data.

15.2 DATA FUSION ARCHITECTURE

The IoT is linked to a broader range of wireless access networks via the interface, Besides access to compatible internet. It comprehends the relationship between articles and object communication and trade and can carry out participated investments with object-to-speech communication. Somewhat has a percentage of visibility. Data transmission is trustworthy, and combining an item to organize a definite amount of data is a smart idea. To pull impressive data from several IoT data, data fusion practices are performed. It conglomerates data from manifold sensory sources to yield consequences that are reliable, precise, and comprehensive. One of the utmost substantial progressions in the web of things is data aggregation. "Combine multiple sources to collect less expensive and more advanced data or more efficient data," Castanedo describes the data combination. Data fusion has been established in a diversity of sensory regions across several domains.

The IoT produces a lot of raw data in its current condition. WSNs that arrange an IoT device detect environmental variables and offer information about employees. On the employee side, this data is also controlled by data mining counts in order to get a general picture of the situation and then make a control action. The information delivered by sensory centers is frequently inaccurate. Before extracting vital data, significant calculations should be conducted on the data. Before the data mining can begin, the calculation must be completed. Data mining is the last step in the data extraction process. However, before the data can be processed in the mines, it must be well-organized in order to extract a limited or unexpected pattern from the data. Fusion data is a variety of data processing that connects and detains data from a number of different sources.

Collaborate on real-time monitoring, viewing, and data collection in the area of various natural or visual texts. Receiving tangible and quantitative items, in addition to people and objects from a broad variety of communication. It is linked to the Internet and serves as a large-scale data processing platform. Prior to the control of object systems, such as sensor organizations, a ton of sensory hub must be passed to the location inside with the purpose of collecting exact information. At the heart of the compact sensor hub, to improve the accuracy of integrated data across the whole system. There will be a definite link of space if posted in high-density locations, in the same object, or in the event of watching domains that come together to receive report data. The transmission, like a comparable harp, is more than simply a passing glance. If there's a likelihood that all hubs will look at the data delivered to the meeting point, the system's transfer speed will slow down to prevent resource waste. Furthermore, many data transfers occurring at identical times can cause a major struggle, lowering book performance. In addition, because data transfer is the key variable in the hub's power consumption, excess data transfer will use the extra energy messengers from the site to shorten the pattern of the scheduled sensor's presence. Data Fusion Architectures can be arranged in a variety of ways.

15.2.1 CENTRALIZED FUSION ARCHITECTURE

The integrated amalgamation structure produces an adequate prediction response when the data is well-organized in relation and there is no limit to the data transport aptitude of the data. Adjusting all of the data in the central hub, alternatively, raises a number of concerns, including the middle class's high computation burden, the high demand for book transfers, the likelihood of failure (owing to central hub failure), and the intensity of architectural modifications. The key connecting point in this engineering is in the central processor wing, and all alternatives were chosen there. Figure 15.1 depicts this architecture.

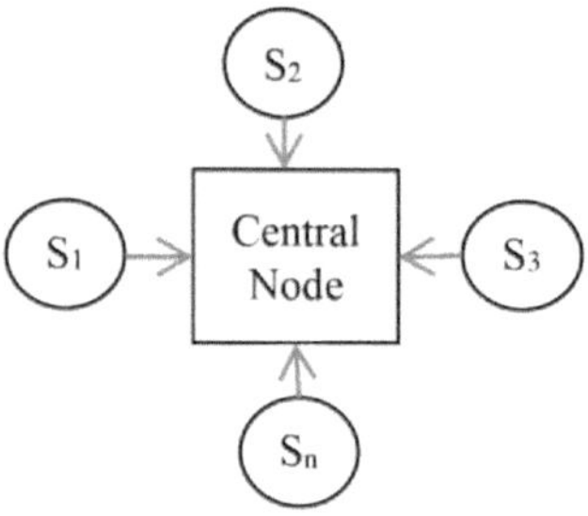

FIGURE 15.1 Centralized fusion architecture.

15.2.2 ARCHITECTURE OF DISTRIBUTED FUSION

The fusion structure, Alternatively, is an isolated area in which the nodes operate unconventionally while sharing information with each other external to the central fusion node. The input from the dispersed process was then integrated with other sensory data by the central processor. This structure can be thought of as a further refined version of central fusion formulation [13]. Figure 15.2 depicts its structure.

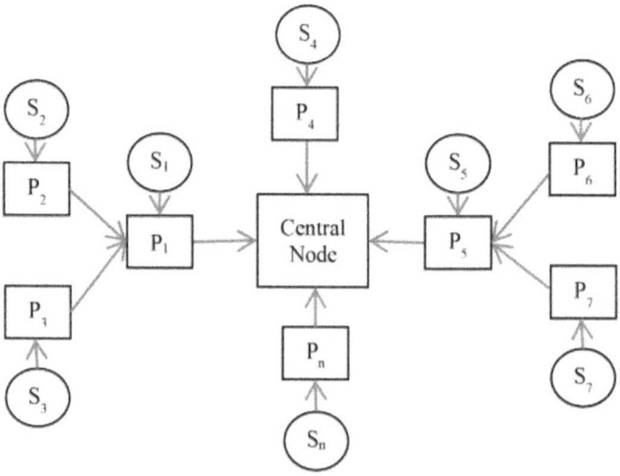

FIGURE 15.2 Disseminated fusion architecture.

15.2.3 HYBRID-FUSION ARCHITECTURE

A novel architecture known as hybrid fusion architecture is generated by integrating two surviving structures. The hybrid architecture has a high level of exactness in a diversity of circumstances and with any sort of data. This structure is tremendously adaptable, and it consents to a range of centralized and distributed designs to be fused and novel hybrid architectures to be created using numerous methods.

15.3 LITERATURE REVIEW

In WSNs, there is a lot of study on IoT data amalgamation systems that learn data amalgamation. To stimulate the IoT, these technologies employ advanced data fusion algorithms such as Classical inference, Dempster-Shafer, Bayesian, and fuzzy logic systems. There are numerous data aggregation processes carried out by various experts. True Procedures are a means of delivering standard and custom-designed methods like fuzzy logic and neural networks (NNs), in addition to Bayesian Methods, which are used to combine multiple sensory inputs. It is necessary to scrutinize the prospects of data consolidation. To reduce the danger of disappointment, a systematic vehicle storage structure based on fuzzy multi-sensor data fusion (MSDF) is being investigated. Central Fusion and Distributed Fusion are the two projects described here.

Based on the structure of three assemblies, Alessandra De Paola offers a self-contained, context-sensitive, flexible system for the aggregation of sensory data [1]. At the intermediate level, heterogeneous data compiled by sensors at a very low level is combined with a unique Bayes network, which also includes sensible data to filter out the disruptive process. The self-assembly technique regenerates the tactile base at the highest level by reviewing a group of nerves to reduce energy expenditure and increase the exactitude of the input. Because of exterior noise and assumed mechanical degradation, the Bayesian methodology allows us to control touch-sensitive balancing [12]. Kaleghale et al. gave a philosophical overview of data analysis. Data aggregation is a disciplinary test field with a comprehensive range of applications expected from regions, including security, mechanical technology, computer use, and an intelligent framework system, in addition to a consent example [4]. This has been and will continue to be the driving force behind the testing network's increased interest in developing advanced development algorithms and models. Other scientific data-driven methods for data amalgamation have been introduced, in addition to observational ideas, linked guessing networks, and statistics that are open to all classes. Besides, the data aggregation network now includes a number of advanced test sites.

With the purpose of developing the perception of employment, Zheng et al. apply a rational strategy to shooting from various sensors. A vague logic technique [5] can show and link images to augment the contrast of the integrated image with the help of vague rules and if membership functions are designed for image data setup. A trend that is used to integrate images is the Mamdani system. Submissive testing findings also demonstrate that it deviates greatly from its intended goal from time to time [6]. Padovitz et al. created a method that is distinguished by the high quality of sensory fusion in a context that is aware of the context [7]. This method is based on the surviving amalgamation process, which can be used for a multiplicity of reasons. The strategy is intended to establish logical applications and provide various areas of interest in data aggregation in a context-aware region. It's a hazy idea of what a state of affairs is: a state of affairs is defined as the slogan of the qualities inside an acceptable Region of the states. They make an effort to test their senses. Paola et al. demonstrate that Bayesian Networks built a useful data management tool before improving artificial intelligence (AI) [9]. The Bayes network model includes a meta-level level that justifies the dynamic structure of the visual framework by making proof of probable assumptions available. The additional meta-level denotes the exactitude of the system's outcome in addition to the expense of utilizing the sensory infrastructure. By modifying the operating system of the fundamental sensory device, the best balance between execution and cost was to be found. The Bayesian network's self-preparation of witnessing sites is accomplished using multi-on-line optimization [10]. The model demonstrated that the proposed strategy has the prospective to overcome the problems provided by the irreversible loss of sensory balance, allowing for a more immediate result while simultaneously lowering expenses associated with energy use.

Zhang et al. offer a data amalgamation system that usages unique Bayesian networks to give the power, power, purpose, and performance of data amalgamation in a timely manner with fewer devices [8]. The transcendental component is the Vigorous Bayesian network as a fusion structure, which provides insight into and fully integrates the growing structure for the display, reconciliation, and acquisition of pertinent data for diverse approaches at innumerable levels of research. The suggested approach is appropriate for situations in which elections should be performed using widely available data from a diversity of sources.

15.4 MULTI-SENSOR FUSION OF DATA

The fusion data module provided a robust Bayes network capable of capturing the intensity of these under-observation scenarios and looking beyond present scenarios to prior countries. There is a diversity of data amalgamation approaches available, including Markov's hidden models [2]. The Vigorous Bayesian network facilitates us to replicate the data that indicates the visible expression of a hidden world power across time. It is a great tool for data aggregation [4].

Here the algorithm is designed for the Data Fusion module. Its main purpose is to include the state of the country, in the custom of a given aspect of interest. Represented by hidden X_t variable. The set of sensory studies compiled by those sensors operating at the time 't' is represented by $E_t = (E_1, \ldots, E_n)$, Conferring to the Optimization Optra module. The context information, i.e., $C_t = (C_1, \ldots, C_k)$, depends largely on the state of the application. The sensor model is defined by the distribution of opportunities $P(E_t X_t)$ representing how sensory learning is affected by the present state of the system. The belief about a specific system state in the time slice 't' is defined with the first-order Markov model as:

$$Bel(xt) = P(xt \mid E1:t, C1:t) \tag{1}$$

To obtain a practical formulation of our belief, Bayes rule is applied, and expressed as follows:

$$Bel(xt) = P(xt \mid E1:t, C1:t) = P(xt \mid E1:t-1, E1, C1:t) \tag{2}$$

$$= \eta \cdot P(Et \mid xt, E1:t-1, C1:t) \cdot P(xt \mid E1:t-1, C1:t) \tag{3}$$

where; 'η' is a normalizing persistent. After this, the Markov assumption, gives the value of the parent node X_t, by assuming that sensor measurements are mutually conditionally independent:

$$P(Et \mid xt, E1:t-1, C1:t) = P(Et \mid xt, C1:t) = P(Et \mid xt) \tag{4}$$

where; E_t is the specific value of the sensory reading gathered by the sensor 'i' in the time slice 't.' Finally, this can be defined with the subsequent recursive formula:

$$Bel(x_t) = \eta \cdot \Pi_{e_t^i} P(e_t^i \mid x_t) \cdot \Sigma_{x_t-1} P(x_t \mid x_{t-1}, C_t) \cdot Bel(x_{t-1}) \tag{5}$$

where; 'α' is integrated with the normalization constant 'η.' Here 'n' is the numeral of sensor nodes and 'm' is the number of conceivable values of X_t. Consequently, the inclusive complexity of computing $Bel(x_t)$ for all values of X_t is $O(m_2 + m \cdot n)$. Using this equation, the inference can be performed, where the time and space required for updating the network belief are self-determining of the sequence length.

15.5 CONCLUSION

In the IoT, data fusion can diminish networking traffic, reduce energy consumption, improve the exactness of the outcomes, and improve network performance discernment. The current implementations of data fusion methods will be discussed in depth in this investigation. Their hierarchical structure will be illuminated, and standard parameters will be linked to it. By means of the upgradation in data fusion methods, additional hierarchical resolution will be provided, and it will be connected with the existing methods. This study will go beyond IoT's hierarchical implementation and cover all layers of data abstraction, from sensor to server.

KEYWORDS

- **Bayes rule**
- **data fusion architecture**
- **internet of things**
- **Markov model**
- **multi-sensor data fusion**
- **real-time data processing**

REFERENCES

1. Alessandra De, P., Pierluca, F., Salvatore, G., Giuseppe Lo Re., & Sajal, K. D., (2016). An adaptive Bayesian system for context-aware data fusion in smart environments. *IEEE Transactions on Mobile Computing*. doi: 10.1109/TMC.2016.2599158.
2. Sanchez, D., Tentori, M., & Favela, J., (2007). Hidden Markov models for activity recognition in ambient intelligence environments. In: *8 Mexican Int'l Conf. on Current Trends in Computer Science* (pp. 33–40). IEEE.
3. Cook, D., Augusto, J., & Jakkula, V., (2009). Ambient intelligence: Technologies, applications, and opportunities. *Pervasive and Mobile Computing, 5*(4), 277–298.
4. Khaleghi, B., Khamis, A., Karray, F. O., & Razavi, S. N., (2013). Multi- sensor data fusion: A review of the state-of-the-art. *Information Fusion, 14*(1), 28–44.
5. Zheng, Y., & Zheng, P., (2010). Multi-sensor image fusion using fuzzy logic for surveillance systems. In: *Proc. 7 Int'l Conf. on Fuzzy Systems and Knowledge Discovery (FSKD), 2*, 588–592.
6. Hall, D., & McMullen, S., (2004). *Mathematical Techniques in Multi-Sensor Data Fusion*. Artech House.
7. Padovitz, A., Loke, S. W., Zaslavsky, A., Burg, B., & Bartolini, C., (2005). An approach to data fusion for context awareness. In: *Modeling and Using Context* (pp. 353–367). Springer.
8. Zhang, Y., & Ji, Q., (2006). Active and dynamic information fusion for multi-sensor systems with dynamic Bayesian networks. *IEEE Trans. Systems, Man, and Cybernetics, Part B: Cybernetics, 36*(2), 467–472.
9. De Paola, A., La Cascia, M., Lo Re, G., Morana, M., & Ortolani, M., (2012). User detection through multi-sensor fusion in an AMI scenario. In: *Proc. 15 Int'l Conf. on Information Fusion (FUSION)* (pp. 2502–2509). IEEE.
10. De Paola, A., Gaglio, S., Lo Re, G., & Ortolani, M., (2011). Multi-sensor fusion through adaptive Bayesian networks. In: *AI*IA 2011: Artificial Intelligence Around Man and Beyond* (pp. 360–371). Springer.
11. Hossain, M. A., Atrey, P. K., & El Saddik, A., (2009). Learning multi-sensor confidence using a reward-and-punishment mechanism. *IEEE Trans. Instrumentation and Measurement, 58*(5), 1525–1534.
12. De Paola, A., & Gagliano, L., (2014). Design of an adaptive Bayesian system for sensor data fusion. In: *Advances onto the Internet of Things* (pp. 61–76). Springer.
13. Makarenko, A., Brooks, A., Kaupp, T., Durrant-Whyte, H. F., & Dellaert, F., (2009). Decentralized data fusion: A graphical model approach. In: *Proc. of the International Conference on Information Fusion*, 545–554.

Index

3

3D virtual
  biomimetic network, 6
  bug catching networks, 6

A

Academic
  agreement, 214
  institution, 117, 123
  records, 123
Accelerometers, 79
Access control
  list, 173
  policies, 128
Accessibility, 1, 6, 58, 125, 140, 244, 250
Accommodation monitoring, 22
Accountability, 129
  department, 123
Accurate
  data management, 129
  measurement, 41
Acoustic audio indicators, 32
Active
  passive sensor, 97
  sensors, 97
Actuators, 2, 12, 44, 98, 135, 160, 162, 220, 222
Adaptability, 6, 36
  intelligent compressive sensing scheme, 21
Administer
  imperatives, 246
  medications, 238
  nodes, 35
  records, 61
  report, 136
Admission
  details, 122
  enrollment, 120
Adobe
  Audition CC, 124
  Premiere Pro, 124
Advanced
  information technology hardware, 218
  message queuing protocol (AMQP), 167, 175
Aeroponic, 230–233
  approach, 230
  farming, 230, 231, 234–236, 238
  medium, 236
  system, 231–234, 241
    nutrition content, 236
Affiliation, 16, 135
Agent-on-chip (AoC), 16
Agribots, 199–201
Agricultural, 1, 2, 5, 23, 53–57, 61, 62, 69, 102, 113, 193–195, 197–199, 201, 202, 205–207, 211, 213–216, 218, 220, 223–226, 229, 236
  aid, 200
  commodities, 213
  countries, 55
  crop, 201
  department, 55
  engineering, 193, 207
  equipment, 217
  field
    conditions, 221
    management, 197
  industries, 54, 55, 199, 206
  investment plans, 215
  land, 63, 195, 196, 198, 201, 202, 213, 230
    registrations, 63
  market, 54, 213
  methods, 59
  opportunity, 214
  policies, 55
  practices, 195, 196, 205–207
  production, 207, 215

products, 64
robot, 194, 199, 200
sector, 193, 200, 225
sites, 201
supply
chain network, 56
chain, 53, 56, 57, 62
process, 57
technology, 218
AgriDigital, 63
Agroforestry, 194
Agronomical activities, 198
Air
conditioner, 7, 259
pollution, 6, 27, 196
monitoring, 27
pressure, 205
quality, 76, 77, 161
Aircraft, 1
Alarm webcam switching system, 262
Alert frameworks, 166
Algorithmic determination, 16
Alignment, 79
Alternative hypothesis, 148
Alumni
association, 121
module, 121
Amalgamation
process, 280
system, 16
Amazon, 83, 85
web services, 83
Analog
digital sensor, 97
domain neuromorphic implementation, 32
to-digital converter, 17
Antennas, 2
Application sector, 23
case studies, 29
energy awareness using wireless sensor networks (WSNS), 34
pedestrian counting-number, 33
sensor nodes (high-temperature-wide temperature), 31
human footstep sound classification, 32
commercial use, 29
environment, 27
healthcare, 25
home automation, 28
military, 23
Appropriate characterization techniques, 236
Aquatic ecosystem, 196
Arabidopsis, 234, 241
Architectural
design, 190
modifications, 278
Arduino
board, 44, 51, 111, 222
Development Environment, 44
microcontroller, 220
UNO Board, 43
Yun circuit interface, 9
Artificial
intelligence (AI), 71, 73, 91, 111, 112, 119, 174, 193, 199, 205, 197, 206, 207, 280
internet GPRS Wi-Fi, 129
mother monitoring system, 71, 72, 85, 88
Asset
arrangement, 6
designation, 245
Assignment submission, 120, 121
Assistance framework, 156
Asthma identification, 26
ATMega88 microcontroller, 11
Atmospheric discharges, 27
Attack detection, 23
Audio
signature instrument, 44
video tools, 124
visuals, 124
Auditing departments, 55
Authenticity, 129
Automated, 197, 205
agronomical practices, 206
data, 99
generated message, 187, 188
integrated information, 128
internet protocols, 265
irrigation systems, 197
machine adjustments, 207
physical systems, 14
services, 29

tasks, 78
technology, 197, 199, 211
tractors, 197
window blind adjustment, 161
Automotive industry, 29
Autonomous tractors, 219
Auto-steering, 201
Auxiliary clients, 244, 246–248
Availability, 129, 147
Average
absolute error (AAE), 146, 151, 152
relative error (ARE), 146, 147, 151, 152

B

Back-propagation algorithm, 144
Barilla, 63
Base station (BS), 6, 7, 9, 13, 19, 22, 225, 244, 249, 250
Bassinet, 82
Battery
free wireless sensor, 14
powered battery WSN hubs, 8
sensor hubs, 1
Bayes rule, 281, 282
Bayesian
methodology, 279
network self-preparation, 280
Biological food poisoning, 235
Biometrics, 79
Biosensors, 79
Blended learning, 130
Blockchain, 53–58, 60–69
IBM cloud infrastructure, 63
ledger, 65, 66
model, 66
network, 58
technology, 53–57, 62, 63, 69
Blood pressure, 4, 25
Bluetooth, 39, 43, 47–49, 51, 82, 85, 99–101, 136, 162, 166, 167, 180, 220
enabled device, 101
HC-06 Module, 43
module, 39, 43, 48, 49, 51
versions, 101
Board, 43, 44, 83, 85, 127, 136, 222, 244, 248, 250, 272
Body temperature, 25, 41, 47, 83
Boolean expressions, 68
Bootstrap framework, 66
Boron, 230
Botnets, 170, 175
Breadboard, 181
Broadband wireless technology, 7
Bug-catching network geography, 6
Building
automation management, 139
energy management systems (BEMSs), 12
Built-in
applications, 276
laptop computers, 100
PIR motion sensor, 84
Bunching conventions, 254
Business
exploration conditions, 156
functionality, 141
layer, 136
methodologies, 136
Buzzer-speaker, 44

C

Cable remote control, 265
Calcium, 230
Camera module, 86, 87
CamStudio, 124
Cancer identification, 26
Canva, 124
Carbon
dioxide (CO2), 6, 33, 82, 212, 215
footprint, 161
Carrier senses multiple accesses with collision avoidance (CSMA-CA), 101
Catastrophe management, 22
Cell phone
application, 159
number, 46
Centigrade, 41
Central
clustering, 249
fusion, 279
formulation, 278
node, 278

interconnected mechanism, 122
management unit, 17
monitoring station (CMS), 7
Centrifugation, 237
Channel
conditions, 246
designation, 248
Characterization, 237
Chemical
contamination, 236
corrosion, 31
fertilizer, 196
industries, 55
food poisoning, 236
Child
abuse, 71
birth, 72
care, 72, 83
checking framework, 81
Chlorine, 230
Choice system, 128
Circuit
boards, 238
breakers, 202
Cisco
Packet, 261, 262, 272
Networking Simulator, 261
Tracer 7.20 networking simulator, 260
Tracer Networking Simulator, 259, 265
Tracer, 259–261, 265, 272
simulator, 261, 262
Clamor sensor, 87
Class attendance, 120, 122
Classical
compressive sensing scheme, 21
inference, 279
sensing scheme, 21
Classification
learning models, 152
web services, 145
Clever
automatic HVAC regulators, 13
remote sensor systems, 6
Clinical consideration gadget, 79
Clostridium botulinum, 235, 241
Cloud
administrations, 226
associated gadget, 179
centric system, 112
computing, 221
platform, 136, 221
database, 261
infrastructure, 225
platform, 222
thing speak web service, 222
Cluster
communication radio, 254
heads (CH), 249
Cobalt, 230
Coffee maker web services, 141
Cognitive radio (CR), 203, 243–245, 247, 248, 250–254
networks (CRNs), 243, 245, 254
sensor networks, 245, 254
Cold storage
centers, 62
companies, 57, 58
Collaborative
learning approach, 127
signal processing, 3
Colossal geological zones, 250
Commercial
business automation techniques, 34
companies, 215
humidity finder, 157
purpose, 29, 30, 36
sector, 1, 22
Communicated, 5–7, 16, 18–21, 25, 28, 43, 45, 64, 77, 98–101, 121, 127, 129, 135, 136, 138, 162, 164, 170, 174, 177, 178, 202, 218–222, 225, 248, 250, 261–263, 265, 276, 277
equipment, 261
information groupings, 246
layer, 220
systems, 218
Community
coverage, 35
deployment, 35
Compact sensor hub, 277
Complex
encoding, 246

nutrient requirements, 231
time-domain capabilities, 32
Component, 42, 51, 173, 183, 280
Composting, 194
Comprehensive applications, 136
Compressive sensing scheme, 21
Computational simulations, 14
Computer
applications, 217
camera, 33
child-rearing condition, 91
network, 95
system, 80, 96
Concentrated solar, 139
Conductivity, 86, 230
Confidential, 129
control access, 129
information, 22, 23, 161, 169, 170
Connected
layer technologies, 178
learning, 139
way frequency grid, 14
Consensus classification decision, 145
Constrained application protocol (CoAP), 99, 165, 175
Consumer services, 102
Contract, 67
Control
monitoring, 161
radio networks, 162
Conventional
acceptable frameworks, 3
power plug, 158
Conversational learning, 127
Cooperative bank, 59
Coordinated electronic innovation, 36
Copper, 230, 261
CorelDRAW, 124
Corporate organization, 173
Correspondence innovation support, 74
Cost, 8, 79, 202
deployment, 23
Cover cropping, 194
Credit system execution, 122
Critical
region, 2, 22, 25, 36
security, 35
Crop
damage, 57
harvesting, 199
management practices, 207
production, 57, 59, 64, 68, 194, 213, 216, 230
rotation, 194, 215
yield, 193–195, 206, 207, 214, 230, 235
Cryptocurrencies, 61
Current graduation details, 123
Customary remote organizations, 247
Custom-designed methods, 279
Customer interactions, 75
Customization approach, 6
Cut up generator, 33
Cyber
attacks, 169, 170, 172
botnets, 170
data theft, 171
IoT smart homes, 169
man-in-middle attacks, 170
remote recording, 171
criminals, 169–171
physical
frameworks detection, 36
systems, 14

D

Daily consumption, 28
Damage proof system, 202
Data
abstraction, 282
aggregation, 136, 198, 276, 279–281
amalgamation, 279–281
analytics solutions, 219
analytics, 218, 219
capability, 10
collection, 20, 88, 99, 244, 277
communications, 261
extraction process, 277
filtering, 136
fusion architecture, 276, 277, 282
architecture (distributed fusion), 278
centralized fusion architecture, 277
hybrid-fusion architecture, 279
fusion methods, 282

fusion module, 281
management tool, 280
packets, 101
sharing, 225
transfer capacity, 246, 247
Database difficulties, 225
DC
DC converter productivity, 8
motors, 98
Decentralized
clustering protocols, 249
ledger, 65, 68
Decision
stump (DS), 33, 146
function vector generator, 33
table (DT), 142, 145, 146, 148, 149, 151, 152
tree, 142, 143
Deep learning, 83, 127, 178
Deforestation, 215
Degree
radiation, 86
temperature strategies, 31
Deionized water, 237
Denial of service attacks, 225
Department module, 120
Dependability, 6
Desalination, 139
Development
autonomous sensing, 5
reasonable bunches, 250
smart homes, 156
Device
micro-processing unit, 16
to-cloud communication, 98
Digital
agricultural, 207, 216–218
solutions, 216
tools, 216
circuit improvements, 226
devices, 211, 259
India, 69
physical structures, 15
programs, 218
technologies, 120
Digitalization, 66, 95
Digitized data robots, 199
Dioxins, 236
Disable universal plug, 173
Disappointment hazard, 6
Discipline, 120
Discretionary customers, 245
Disease
diagnostics, 199
free
cultivation, 230
potato seeds, 230
Disorder observation, 5
Distance education, 123
Distributed
agent computing, 15
event set-membership filtering, 12
fusion, 279
intelligent hungarian algorithm, 10
ledger system, 61
material-implanted frameworks, 15
network, 58
nodes, 35
Distributors, 57, 58, 66
District cooling, 139
Diverse grouping methods, 254
DNA microarrays, 238
Documental proofs, 55
Domestic veggie consumption, 236
Door unlock module, 182
Drastic population growth, 193
Drought
prone areas, 232
stricken areas, 231
Dual-band wireless LAN compliance certification, 85
Dunnett test, 149
Dynamic, 6
host configuration protocol, 265
range access, 248
spectrum, 243

E

Eco-friendly tracking, 22
Ecological disruption, 230
Economic
farming, 197
intelligent gadgets, 36
setup (instrumentation), 3

Edpuzzle, 126
Education, 102, 118–121, 125, 126, 128, 130
  opportunities, 125
  system, 118, 119, 121, 128, 130
Efficient spectrum sensing, 243
E-learning content development, 124
Electrical
  conductance, 231
  devices, 219, 260, 261
  gadget, 179
    surveillance control framework, 179
  signals, 79, 86, 98
  substations, 203
  wastage, 190
Electrochemical, 8, 11, 79
  sensors, 203
Electromagnetic, 99, 100
  fields, 100
  level detection system, 134
  radio signals, 100
  waves, 99
Electromechanical, 44
Electronic
  appliances status module, 182
  assets, 139
  components, 236
  devices, 77, 97, 133, 166, 184, 260, 262, 263, 272
  gadget, 83, 159
  health record (EHR), 112, 113
  sensor, 85
  signage, 173
  smart instruments, 217
  voting, 61
Emergency room, 46
Employment opportunities, 206
Encoding techniques, 246
Energy
  aware routing, 253
  conservation, 197
  consumption, 2, 5, 19, 161, 174, 225, 230, 250, 251, 282
  efficiency, 2
  managing solutions, 27
  mindful steering plans, 249
Engineering-manufacturing, 102
Enhanced
  blended classroom teaching, 117
  diversity, 215
Ensemble
  feature selection technique, 146
  learning scheme, 146
  technique learning model, 149
Entryways, 164
Environment, 233
  crisis, 27
  impact, 161
  observation forecast, 22
  parameters, 219, 222
  protection, 197
  quality, 206
  stress, 216
Equipment
  circuit, 7
  sensors, 18
Eradicating weeds, 232
Escherichia coli microbes, 9
Essential client
  execution debasement, 246
  signals, 246
Ethanol, 237
Ethernet communication, 221
Euclidean distance, 145
Eutrophication, 196, 234, 240
Evaluation-focused ellipsoidal districts, 12
Event management system, 122
Examination
  module, 120
  software monitor, 122
Exponential pattern, 212
Extensible terminology (XML), 226
Ezvid, 124

F

Fabrication, 229, 237
  technique, 238
Face recognition, 178, 179, 182, 184, 186, 190
  module, 182
  technology, 178
Fake transactions, 69
Farmers, 53–61, 63–66, 68, 69, 193, 195, 197–199, 201, 202, 204–206, 211, 215, 216, 218, 221, 225, 226, 232

Farming
  corporations, 204
  domain marine tracking, 23
  related
    applications, 221
    problems, 232
Fault tolerance, 23, 53, 56, 58, 61
Federal Communication Commission (FCC), 244, 245
Fee payment, 120
Fertility, 211
  prescriptions, 218
Fertilization schedules, 217
Fertilizer, 59, 62, 196, 230, 241
  application instruments, 203
  calculator, 199
  treatments, 233
Field
  by-field basis, 217
  health photography, 217
  of view, 33, 85
  View Cab app, 217
Filmora Scrn, 124
Finance
  accounting module, 120
  inclusion, 216
Fire indicator, 156
Firmware vulnerabilities, 169
First-order Markov model, 281
Flexibility, 6, 16
Flow chart, 136, 238
F-measure value, 148, 151
Focal cooling control framework, 7
Food
  contamination, 235
  Corporation of India, 65
  cultivation, 236
  integrity, 57
  poisoning, 235, 236
  production methods, 236
  protection, 193
  public distribution, 65
  safety, 57, 66
  security, 57, 193, 215, 216
  supply chain, 56–58, 62–64, 66
Forest fires monitoring, 27
Framework design, 223
Fraud
  clone website, 66
  detection, 69
Free screen video recorder, 124
FreeCam, 124
Front-end web application, 66
Fruitful grouping frameworks, 250
FT-Raman, 238
Full
  automated robots, 199
  proof effectiveness, 200
Function
  classifiers, 144
  learning models, 144
Fuzzy logic systems, 279

G

Gateway, 80, 98
Gathering information, 23
Generalized exemplars, 145
Glass thermometer, 78
Global
  classrooms, 125
  climate change, 212, 214
  population, 211, 213
  supply chain of crops, 205
  temperatures, 214
  traders-farmers, 54
  warming, 195
    groundwater levels, 196
Globalization, 54
Goals, 79
Google slides, 124
Government, 22, 53–60, 63, 64, 68, 69, 206, 212, 223
  portal, 58
  regulatory authorities, 60
  supported agencies, 56
Gradient descent (GD), 144, 146, 148, 149, 151, 152
Gram Panchayat, 59
Graphical user interface (GUI), 46
Green
  energy initiatives, 161
  house
    monitoring, 27
    products, 199
    tracking, 22
  revolution, 196

Gridlocks, 139
Groceries, 58
  store, 57, 62
Grove framework, 86
GSM module, 45, 48, 87
Guardians, 74, 81
Guidance systems, 204

H

Habitation, 85
Half-breed inhabitancy estimation calculation, 13
Hazardous chemical plants, 35
Health
  care, 4, 5, 22, 25, 26, 36, 39, 102, 112, 113, 140, 174
    industry, 174
    personnel, 25, 26
    sector, 4, 25, 26, 174
  monitoring, 25
  sector, 1
Heart
  beat
    generator circuit, 83
    rate, 4, 25, 82
  rate
    analysis, 47
    sensor, 41, 51
Heating ventilating air conditioning (HVAC), 12–14
Heavy metals, 235
Herbicides, 196, 200, 203
Heterogeneous
  data, 279
  environments, 119
  network, 20
  sensor
    arrangement, 14
    determination, 14
    networks, 19
  systems, 151
Hierarchical structure, 282
High effect pollutants, 9
High income employment generation, 197
High quality
  online education, 117
  screencasts, 126
High resolution satellite images, 137
High sensitivity water sensor, 86
High tech industry, 199
High voltage
  electrical devices, 194, 199
  equipment, 202, 203
Highest productivity, 197
Hijacking assaults, 225
Histogram of oriented gradients (HOG), 179
Home
  appliances, 97, 102
  automation, 1, 3, 22, 23, 28, 75, 99, 113, 161, 180
  computerization, 166
  gateway, 272
  robotization frameworks, 86
Homogeneous
  network, 19
  sensor networks, 19
Horticultural, 5
Host configuration protocol server, 265
Household chores, 73, 90
Human
  intervention, 2, 22, 78, 134, 178, 219
  resources (HRs), 123, 130, 226
  to-machine communication, 74
Humidity, 157, 205
  finder, 157
Hungarian technique, 11
Hybrid fusion architecture, 279
Hydro-atomized nutrient solution, 232
Hydrological, 194
  cycle, 196
Hydroponics, 231, 235

I

Idle listening, 17
IF neutron array, 33
If-then-else statements, 145
Image sensors, 79
Immutability, 58
IMovie, 124
Improved
  accuracy, 197
  comfort, 161
Indian
  Agriculture Policy, 54
  social norms, 72

Indigenous farmers, 194
Individual advanced associates, 82
Indoor reconnaissance, 33
Industrial
control, 23
monitoring, 23
supply chains, 61
Industrialization, 195, 196
Infant, 74, 81–83, 85, 87, 89, 90
Information
communication technologies (ICTs), 137, 140, 152
preparation, 15
transfer technology, 167
Infrared
radiation, 86
transmitting subject, 33
Infrastructures, 53, 133, 134, 137, 138, 226
Inhabitancy estimation time, 13
Inkjet printing technique, 238, 240
Innovative
digital devices, 125
interdisciplinary approaches, 193
IoT smart applications
example of, 102
smart garbage separator, 102
smart healthcare, 111
Insect intrusion, 204
Insecticides, 196
Instance learners, 145
Instructional supplies, 139
Insurance
agencies, 55
companies, 57, 58, 205
Integrated
development environment (IDE), 44, 47, 51, 222
intelligent information, 120
interdisciplinary technology, 193
Integrity, 66, 129
Intellectual
radio, 244
surroundings, 1
Intelligent
agricultural
sector, 197
system, 193, 194, 197, 206, 218, 227
algorithms, 161
ambulance system, 46
applications, 95
artificial mother monitoring system, 72
automatic security system, 178
environment, 36
framework system, 279
home (IH), 133, 134, 136, 140–142, 152, 155
application, 141, 152
realization, 133, 140
IoT management system, 225
model, 120
department module, 122
examination module, 122
finance accounting module, 123
inventory library module, 123
student information module, 121
student placement HR module, 123
networks, 23
node system, 16
nursing homes, 25
parenting system, 71
placement system, 123
regulator, 13
screening, 231
security system, 177, 178, 190
sensor, 3, 4, 19, 23, 25–28
node (ISN), 1–5, 12, 16–19, 21–23, 25, 29, 36, 39, 53, 71, 95, 117, 133, 152, 155, 177, 193, 211, 229, 243, 259, 275
node framework, 22
node system, 2, 3, 18, 21, 35
system, 4, 17, 18
technology, 1, 2, 22, 27
surveillance system (ISS), 118, 119, 130
system
agriculture, 199
for agriculture (IAS), 193, 194, 197, 201, 204–207, 218
technology, 2, 4, 22
transduction sensors, 36
transportation system (ITS), 65, 118, 130, 139
University Management System (i-UMS), 117–123, 128–130
ventilation, 23

Interactivity, 126
Intercropping, 194
Interdisciplinary research, 193, 207
Interlace, 246
Intermediaries, 58, 61
International institutions, 55
Internet, 10, 15, 69, 80, 91, 96, 134, 136, 137, 141, 152, 175, 177, 190, 207, 227, 260, 263, 265, 266, 272, 275–277
  connection, 66, 77, 95, 96, 134
  gateways, 136
  internet of
medical things, 9
    actuators, 98
    applications, 136, 179, 225
    automated agriculture works, 225
    control platforms, 276
    device administration, 225
    elements of, 97
    enabled tools, 276
    end user-devices, 98
    farming (smart) system, 219
    gateway, 98
    infrastructure, 140–142, 225
    network, 95–98, 112, 113
    protocols, 98, 164
    sensors, 97
    smart applications, 113
    smart homes, 160, 169
    technology, 59, 71, 72, 168
    things (IoT), 8, 12, 13, 33, 39, 54, 59, 69, 71–78, 80–83, 90, 95–98, 101, 102, 111–113, 133–136, 140–142, 151, 152, 155, 156, 160–162, 164–175, 177–180, 190, 198, 203, 206, 211, 218–221, 224–227, 259, 260, 262–264, 266, 267, 271, 275–277, 279, 282
  technology, 211
Interoperability
  intellectual radio, 243
  policies, 226
Inventory
  applications, 100
  library module, 120
Investment opportunities, 216
IPaq sensor node, 2
Iron, 230
Irreversible biodiversity losses, 215
Irrigation, 194, 195, 198, 199, 201, 202, 204, 213, 215, 218, 220, 226
Italian multinational food company, 63

J

Java
  language, 261
  Script object notation (JSON), 226, 227
Job
  opportunities, 122
  placement, 123
Jumper wires, 181

K

Kappa statistics, 146, 148, 149, 151
K-dimensional binary vector, 146
Kidnapping, 71
Knowledge construction, 127

L

Land
  investments, 215
  ownership authority, 63
  reform department, 55
  slide detection, 23, 27
Laptop devices, 127
Large-scale
  aeroponic farm, 234
  organizations, 8
Lattice estimation models, 14
Leakage detection, 139
Learning management system, 124
Led lights, 181
Level sensors, 97
Lifestyle technologies, 75
Light
  initiation, 79
  points, 166
  regulation, 23
  sensor, 76, 81
  works VSDC free video editor, 124
Linked guessing networks, 280
Lithography techniques, 238
Liveness detection module, 182

Local
authorities, 57, 58
binary pattern histograms (LBPH), 179
governing bodies, 57
officers, 58
regulatory authority, 59
Logistics operations, 75
Long-Range Wide Area Network (Lora WAN), 15, 33
Low-powered microcontrollers, 5
LPG leakage module, 182

M

Machine
learning, 71, 73, 91, 197, 206, 207, 226
performance, 75
Macintosh conventions, 247
Magnesium, 230
Magnetic field, 100
Mail generation, 182
module, 182
Mamdani system, 280
Manganese, 230
Man-in-middle attacks, 170
Manual data collection, 123
Markov model, 282
Material
embedded sensor network systems, 15
implanted frameworks, 15
Mechanical
sensors, 203
smart machines, 74
systems, 5
technology, 279
Media focuses, 173
Mediator adaptor services, 226, 227
Medical data access, 26
Medium access control (MAC), 3, 36, 247
Memory unit, 17
Message queuing telemetry transport (MQTT), 45, 99, 165, 166, 175
Metal
ball configuration, 81
nanoparticle, 235, 238
Meteorological department, 55
Metropolitan territorial applications, 33
Microcontroller, 43, 45–47, 80, 83, 102, 220–222
board like Arduino Uno, 111
Micro-electromechanical
sensor innovation, 79
systems (MEMS), 5
Micro-irrigation
monitoring system, 202
system, 201, 202
techniques, 201, 202
Microphone, 86
Microsoft
PowerPoint, 124
Whiteboard, 127
Microsystems, 8
Microwave-radar innovation, 79
Middlemen, 58, 61, 64
Middleware layer, 136
Military, 1–3, 5, 22–24, 36
detection, 5
Mineral mapping, 197
Minimum required difference (MRD), 148
Ministry of Consumer Affairs, 65
Misting devices, 234
Mobile
airborne sensor locations, 7
application, 112, 161, 211
phones, 75
stations (MSs), 7
Modernization, 193, 226
Moisture
detecting doors, 102
sensors, 76, 97
Molybdenum, 230
Monitor, 2–4, 7, 9, 12, 13, 21, 23, 25–28, 32, 34, 46, 50, 72, 73, 83, 86, 89, 91, 117, 120, 128, 139, 194, 199, 202–205, 207, 217, 219–222, 234–236, 244, 245, 261, 277
aquatic-wastewater, 23
soil, 220, 227
system, 7, 25, 34, 46, 72, 73, 89, 91, 203, 204
Motion
detector, 79, 85, 86, 266
sensors, 158
Movement sensor, 158

Multi-agent
frameworks, 15
structures, 15
Multifaceted nature, 135
Multi-hop data transmission scheme, 244
Multilingual texting tool, 127
Multimedia presentations, 127
Multi-on-line optimization, 280
Multi-participant commodity management platform, 63
Multiple
geographic regions, 226
IoT devices, 226
stakeholders, 61
Multi-sensor
data fusion (MSDF), 279, 282
independent remote hub, 9
wireless node, 9
Multispectral photo cameras, 200

N

Nano-sensor, 6, 236, 238, 240, 241
gadget, 240
radiation, 6
NASA Sensor Webs Project, 4
Natural
condition, 78
disaster, 29
fighting operator recognition, 79
growing process, 230
resistance, 196
Near-field communication (NFC), 100, 101, 113
enabled devices, 100
Network, 2, 4–6, 9, 12, 14, 16, 18–20, 35, 58, 68, 74, 78–80, 96–100, 112, 113, 128, 134, 136, 140, 155, 162–164, 166, 169, 172, 173, 178–180, 202, 217, 245, 248, 250–252, 254, 260, 261, 263, 265–267, 275, 276, 279–282
associated tickers, 173
devices (IoT system), 136, 162
bus topology, 163
mesh topology, 164
point-to-point topology, 162
star topology, 162
topologies of, 162
environments, 4
layer, 135
lifetime, 250, 254
topology, 162
Neural network (NN), 142, 144, 146, 148, 149, 151, 152, 279
Newborn child-checking framework, 82
Nitrates, 196
Nitrogen, 218, 230
Noise sensors, 72
Non-covering groups, 249
Non-fungible tokens, 61
Non-government agencies, 55
Non-interoperability, 140
Non-nested generalized exemplars (NNge), 145, 146, 148, 149, 151, 152
Non-parametric test, 147
Non-repudiation, 129
Notification mechanism, 174
Novel hybrid architectures, 279
Nuclear power plants, 2, 22
Null hypotheses, 148
Nutrient
medium, 238
spray interval, 233

O

Objective-oriented teaching-learning process, 127
Object-to-speech communication, 276
Observation
capacity territories, 33
frameworks, 166
link stream, 14
Occasion correspondence system, 12
Ocenaudio, 124
Oilseeds, 215
Online
admission, 122
software system, 122
auctions, 61
class sessions, 127
collaborative learning, 127
technology, 127
content development, 125
donations, 61

education, 118, 127
learning, 123
process, 124
project, 121
assignment submission, 121
quizzes, 127
services, 180
submission mechanism, 121
surveillance, 88
teaching, 124, 127, 128
tools, 125, 126
Operational efficiency, 129
Operator behavior, 16
Optical
in-pipe sensors, 8
sensors, 97
Optimization
optra module, 281
spending, 161
Optional beneficiaries, 247
Orchid fruit, 201
Organic electrical devices, 238
Organization
associated lighting, 173
execution, 253
lifetime, 245
policies, 225
vehicular traffic, 14
Organized electronic innovation, 36
Origin-destination (O-D), 14
lattice estimation, 14
system grid, 14
Oscillatory movement, 83
Outcome learning process, 127
Outdoor environment, 96
Overcome climatic challenges, 197
Overheating electric wiring, 33
Overvoltage protection, 203

P

Parallel plate capacitor, 31
Parcel conveyance, 250
Parks maintenance, 139
Passive
infrared (PIR), 13, 33, 34, 72, 82, 85–87
motion sensor, 85
sensor, 33, 34, 72, 85, 87
NFC device, 100
supervision, 32
Pathogen
bacteria, 235
free seed stocks, 233
infected plants, 233
PC organizations, 162
Pedagogy, 122, 127
Pedestrian counting-number, 29
Pension funds, 215
Performance degradation, 243
Person identification system, 140
Personal
area network (PAN), 99, 101, 113, 166
computer, 80, 166
Pesticides, 56, 59, 62, 196, 201, 236
Petri net symbolic handling, 16
PhoneControlled security lock systems, 180
Phosphates, 196
Phosphorus, 218, 230
Photo electronically dependent, 33
Physical
digital structures, 15
interference, 225
sensor nodes, 5
vapor deposition, 238
world occasions, 3
Pi camera, 82, 86
Pinnacle studio, 124
Plant
absorption, 234, 238
development, 230, 232, 236
growth, 200, 216, 230, 241
production, 231
Playhouse flaps, 219
Point-to-point connection, 162
Policymakers, 216
Political considerations, 212
Pollution
accumulation, 196
free agricultural practices, 207
Portability
distribution protocol, 165
highlights, 16
Positioning technologies, 218
Posture detection, 49
Potassium, 218, 230

Potato seeds, 230, 235
Power
consumption, 7, 11, 166, 203, 277
efficiency-consumption node mobility, 35
supply unit, 17
utilization, 165
Practical demonstrations, 128
Pre-characterized boundaries design, 160
Precision, 78, 147, 216
agriculture plans, 218
farming, 213, 216
Pregnancy, 72
Presentation tools, 124
Present-day sensor nodes, 5
Pressure sensors, 97
Prezi, 124
Primary users (PU), 243–245, 250
free channels, 243
Printed
electrical devices, 238
gadgets techniques, 9
Private sectors, 1
Processing, 1–3, 6, 17, 19–21, 32, 63, 74, 98, 99, 112, 128, 141, 178, 180, 182, 184, 214, 215, 219, 235, 277
unit, 17
Profitability improvement, 139
Programed
configurable radio, 243
IoT child checking framework, 74
Progress monitoring, 122
Projeqt, 127
Proposed model, 47, 57, 218
farmers, 59
government, 57
system design, 219
traders, 60
Protocol, 98, 99, 164
sensor systems, 3
Proximity sensors, 97
Psychological
radio, 246, 248, 251
organization ideal models, 246
technologies, 260
Public
blockchain, 58, 67
infrastructure investments, 216
spending, 215
Pushbuttons, 181
Python, 178, 190, 261, 262, 266
languages, 190

Q

Quality
administration bunching, 254
monitoring, 139
QoL (quality of life), 137, 138, 140, 152
QoS (quality of service), 4, 26, 140–142, 146, 147, 254
parameters, 142, 146, 147
requirements, 142
Quantitative items, 277
Quantum cryptographic techniques, 9

R

Radio
environment, 245
frequency
identification (RFID), 45, 46, 51, 99–101, 123, 139, 178
transmitters, 46
nuclides, 236
Random
decision forest (RDF), 142, 143, 146, 148, 149, 151, 152
neural network (RNN), 12, 13
inhabitancy estimator, 13
Raspberry Pi, 82, 84–86, 178–180, 190
3 Model B+, 85
robot, 86
speaker, 86
Raw sensory data, 1
Real-time
applications, 112, 113
data processing, 178, 282
monitoring, 8
Reasonably priced, 233
Receive mail response, 182
Reception, 2, 17, 22, 47
Reconciliation, 10, 281
Reconfigurable pipeline conveying process, 16
Record land registry, 63
Reduced root-to-root contact, 232

Refreshed data innovation-based venture, 6
Regular education mechanisms, 126
Related terminologies, 5, 74
  Internet of things (IoT), 74
    connection, 76
    aspects, 77
    power, 75
    why IoT matter, 75
  sensors, 78
    classification, 79
    criteria, 78
Remind, 127
Remote
  door association, 139
  sensing system, 202
  sensor systems, 1
Renewable energy, 234
Repeatability, 79
Reporting station (RS), 7
Research
  cantered approach, 118
  communities, 193
  conducted measurements, 46
Reshuffling clustered compressive sensing, 21
Resilient intelligent wireless sensor networks, 6
Resistors, 31, 181
Resource-limited sensor nodes, 32
Result-oriented information, 117
Robo chef, 174
Robotics, 219
Routing, 136
Runtime topologies management, 35
Rural infrastructure, 214

S

Salinity mapping, 204
Salmonella spp., 235
Sanitizing machine, 46
Sans battery remote sensor, 15
Satellite framework, 247
Savvy comfort-detecting system, 11
Scholarship, 120
Scientific data-driven methods, 280
Screen recorder screen capture, 124
Screencast-O-matic, 126
SD card, 81, 181
Secondary
  nutrients, 230
  users, 245
Security, 6, 28, 29, 36, 55, 58, 61, 64, 71, 79, 85, 90, 91, 119, 128, 129, 140, 155, 158, 159, 161, 169, 171–174, 177–180, 183, 184, 190, 225, 249, 250, 260, 266, 279
  cameras, 158
  framework, 179
  recognition, 79
  surveillance, 29
Self
  assembly technique, 279
  detector, 29
Semantic, 225, 226
  interoperability, 226
Semiconductor
  ATmega328P microcontroller, 43
  materials, 236
Sending SMS, 49
Sensing
  element, 16
  mechanism, 35
  scheme, 20, 21, 36
  technology, 218
  unit, 17
Sensor, 1–20, 22–27, 29, 31–36, 40–42, 46–49, 51, 69, 72, 76–83, 85–88, 91, 96–99, 102, 111, 113, 118, 129, 134–136, 139, 156–162, 164, 166, 168, 169, 174, 175, 178–182, 184, 190, 199, 202–204, 211, 216, 217, 219–223, 225, 226, 229, 231, 235, 236, 240, 243–245, 249–253, 260, 262, 272, 275, 277, 279–282
  data, 6, 14, 80, 174, 175, 226, 253
  frameworks, 36, 82
  gadget, 5, 240
  hubs, 3, 5, 6, 8, 10, 12, 13, 22, 33, 252
  innovative capability, 17
  layer, 219
  learning capability, 17
  modules, 48
  network, 2, 19, 23, 26, 35, 245
    heterogeneous, 20

homogeneous, 19
technology, 46
types of, 19
node, 4, 5, 17, 19, 2, 290, 32, 91, 99, 203, 243, 244, 282
platform, 2
technology, 5
orchestration, 3
organization, 8, 83, 166, 245, 251, 277
approach, 8
packages, 35
technology, 1, 20, 22, 23, 26, 27, 29, 217
units-nodes (SU), 6, 146, 147, 244, 245
Sensory
devices, 260
fusion, 280
information, 4, 22, 28
infrastructure, 280
technology, 2
Service administration, 136
Servo motors, 98, 102, 111
Session hijacking, 225
ShareX, 124
Sheskin multiple comparison test, 151
Shipping
companies, 58
costs, 75
Short-distance electricity transmission, 31
Short-range communication, 100, 136
Shrewd city, 137
Signal enhancement, 36
Simple tuple-space database, 16
Simulation examination, 32
Single digital ledger, 58
Sintering methods, 31
Skin illnesses, 230
Sleep, 17, 72, 266
Slice
generator, 33
presentations, 124
Small remote sensing devices, 198
Small-sized static sensor nodes, 32
Smart
agricultural
intelligent agricultural system, 205
students, 206
system, 203
technologies irrigation system, 201
application, 102, 111, 140, 152
autonomous living, 136
buildings, 139
cameras, 170, 174, 178, 211
cell, 272
chair
system, 42
technology, 50
cities, 137
conceptualization, 138
smart features, 138
contract, 58, 67, 68
digital technology, 261
dust project, 4
effective implementations, 174
electrical appliances, 141
energy, 139
farming, 218, 219, 224
fridge, 174
glasses, 136
grid, 1, 139, 161
home, 155, 168, 172
devices, 170
organization, 158
indoor regulator, 158
irrigation (public spaces), 139
judgments, 276
lighting, 139
system, 134
monitoring, 139
parenting, 71, 91
parking, 139
systems, 134
phones, 75, 100, 122
plugs, 158
refrigerator, 161
roads, 134
security bell, 159
storage, 139
technological equipment, 203
traffic management web service, 141, 142
transportation, 136
vehicle, 136
watches, 100, 171
water, 139
Smoke detector, 184, 190
module, 182

Social insurance check, 5
Socioeconomic costs, 213
Socrative, 127
Soft commodity trading organizations, 214
Software, 32, 49, 80, 90, 96, 98, 117, 119, 120, 123, 124, 127, 130, 141, 197, 198, 216, 217, 226, 262
  application, 117, 123, 216, 218
  development, 119, 120, 141
Soil
  air porousness measurement, 203
  erosion, 197, 202
  health department, 55
  moisture sensor, 201, 220, 222
  salinization problems, 195
  science, 194
  study, 198
  testing mapping, 197
Solar
  cells, 238
  energy harvesting wireless sensor network (SEH-WSN), 8
  panel, 266
    efficiency, 8
  PV, 139
Solution techniques, 238
Sophisticated remote sensor system, 15
Sound
  sensor module Raspberry Pi, 86
  system, 262
Soybeans, 215
Space time issues, 2
Spacecraft, 1
Spatial domain, 21
Spatiotemporally rich data, 8
Speakerphone, 166
Speakers, 86
Specialized
  agricultural applications, 198
  designed robotic arms, 201
Specific
  blockchain platform, 67
  write-in agricultural area, 201
Spectrum
  detecting model, 252
  sensing, 254
Spring-stacked engine, 83
Staff recruitment, 123
Standard agricultural practices, 195
State-of-art manufacturing plants, 23
Statistical surveys, 8
Status-report-service-report, 7
Stock storage materials, 123
Stratosphere, 260
Structural
  health monitoring
    applications, 14
    system, 134
  organization, 119
  patient care technique, 4
Student
  attendance, 122
  information module, 120
  placement-HR module, 120
Subject
  allotment, 122
  management, 122
Substantial progressions, 276
Sulfur, 230
Sun
  oriented fueled battery-accusing arrangement, 8
  powered photovoltaic vitality, 8
Supercomputer, 75
Supermarkets, 57
Supervisory control, 139
Supply
  chain, 55, 57, 63, 66, 75, 96, 215, 216, 235
  demand imbalance, 215
Support
  policies, 216
  vector machine (SVM), 146
Surface acoustic wave (SAW), 31
Surpassed inflation, 215
Surveillance
  cameras, 173
  program, 178
Sustainable
  intensification, 215
  use (natural resources), 197
Switchgear, 202
Swung daintily, 74
Syntactic, 225
  interoperability, 226

Synthesis, 229, 237
System
  administration innovation, 245
  disappointments, 168
  health control, 23
  monitoring air pollution, 6
  smart chair, 50
Systematic
  disposal (waste), 102
  training, 125
  vehicle storage structure, 279

T

TalkingPoints, 127
Teaching
  learning, 122
  staff, 122
Techniques, 237
  fabricating organic devices, 238
  synthesis analysis, 237
Technological innovations, 138
Tele-care, 25
Temperate trimming, 226
Temperature, 4, 9, 13, 15, 17, 23, 28, 29, 31, 33, 39–41, 46–49, 51, 76–78, 82, 86, 88, 97, 140, 141, 158, 169, 173, 174, 198, 202, 205, 211, 214, 218–223, 230, 237
  calculation, 46
  sensor, 41, 48, 86, 97
Text
  editors, 124
  message generation module, 182
Time-basic applications, 36
TinyTake, 124
Topography boundaries, 203
Total volatile organic portion (TVOC), 33
Tower farms systems, 232
Traceability, 62–64
Track humidity, 161
Traders, 53–58, 60, 64, 66, 68, 69, 215
Trading companies, 60
Traditional
  agriculture, 193–197, 207, 230
    erosion, 196
    fertilizers eutrophication, 196
    loss (agricultural land), 195
    management, 194
    pesticide resistance, 196
    pollution-silt, 196
    practices, 195
    soil salinization, 195
    water depletion, 195
  classrooms, 125
  conventional agricultural techniques, 194
  education system, 119
  farming
    re-o, 197, 200rientation, 197
  irrigation strategy, 202
  knowledge, 194
  methods, 238
  paper ledger system, 64
  sensor, 5
  vehicle operations, 204
Traffic
  congestion, 134
  data, 14
  management service, 141, 142
  observation, 5
  security observation, 79
  signal management, 45
  solutions, 139
  superintendents, 139
Transceiver unit, 17
Transformers, 202
Transmission, 2, 9–11, 15, 17, 18, 22, 23, 31, 43, 45, 47, 99, 136, 164, 232, 248, 251, 260, 276, 277
  media, 136
Transparency, 57, 58, 61–66, 161
Transportation, 102, 134, 138
  agencies, 57
Tree classifier, 146
Trustable channel, 57

U

Ubuntu Linux Operating system, 119
Ultrasonic sensor, 39, 42, 46–49, 51, 111, 180, 183
Universal
  intelligence system, 118
  ledger, 53, 56, 62
  management, 119, 120, 125
    software, 120

monitoring system, 117, 120–122, 129
Unmanned aerial vehicle (UAV), 2, 3
Unpredictability, 36, 163
Untrusted cloud platform, 179
Unwavering
efficiency, 11
quality, 6, 15
Urban
fog figuring-dependent sensor-cloud frameworks (SCS-UFC), 10
households, 206
noise maps, 134
territories, 136
Urbanization, 137, 195, 196, 214
Utilization, 251

V

Variable spraying, 203
Vehicle
dynamic frameworks, 79
tracking, 46
Versatility, 6
Video
camera, 166
games, 79
surveillance system (motion detection method), 82
Vinyl tape, 41
Virtual
medical evaluations, 25
private organization (VPN), 172, 175
string thickness, 6
Virtualization facility, 112
Visme, 124
Visualization, 161, 223
Vitality-productive manner, 10

W

Walk speed assessment, 33
Warning systems, 44
Waste management, 134, 139
system, 134
Water
automation system, 139
conservation, 201, 232
consumption, 202, 232
instigated contact, 86
management, 202, 206, 207
nutrient environment, 230
requirement analysis, 199
sensor module, 86
Watershed monitoring systems, 27
Weather
forecasting, 197, 206
service, 141, 142
Weatherproof docking, 200
Web
based instructions (WBI), 123, 130
service
classification, 141, 145–149, 151, 152
classifier, 145
selection, 140, 152
technology, 133, 141, 152
Weed
mapping, 203
measurement, 201
Wi-Fi
associated microcontroller managed system, 180
passive temperature sensor, 31
Wireless, 3, 5–7, 9, 17, 25, 31, 34, 43, 85, 96, 99–101, 113, 134, 136, 164, 166, 179, 180, 202, 203, 243, 244, 248, 250–252, 260–263, 265, 266, 268, 272, 276
body area networks, 26
communication technique, 100
fidelity, 100, 101
intelligent
GPS sensor network (WINGSNET), 6, 7
sensor actuator network (WISAN), 3
networking equipment, 262
router, 100, 263, 265, 266, 268, 272
sensor network (WSN), 4, 5, 7, 8, 10, 12, 29, 32, 35, 202, 243–245, 248–250, 277, 279
Workload balance (Sensor-Cloud Systems), 10
World
boxing system, 199
Wide Web, 125

Y

Yield monitoring frameworks, 203

Z

ZigBee, 9, 10, 136, 166, 218
highlight point remote shrewd sensor device, 10
knowledge transmission, 10
Zinc, 230